Speaker's Corner Books

is a provocative new series designed to stimulate, educate, and foster discussion on significant public policy topics. Written by experts in a variety of fields, these brief and engaging books should be read by anyone interested in the trends and issues that shape our society.

More thought-provoking titles
in the Speaker's Corner series

Two Wands, One Nation
An Essay on Race and Community in America
Richard D. Lamm

The Enduring Wilderness:
Protecting Our Natural Heritage through the Wilderness Act
Doug Scott

Parting Shots from My Brittle Bow:
Reflections on American Politics and Life
Eugene J. McCarthy

The Brave New World of Health Care
Richard D. Lamm

Social Security and the Golden Age:
An Essay on the New American Demographic
George McGovern

For more information, visit our Web site,
www.fulcrum-books.com

Think for Yourself!

an Essay on Cutting through the Babble, the Bias, and the Hype

Steve Hindes

Fulcrum Publishing

Golden, Colorado

Library of Congress Cataloging-in-Publication Data

Hindes, Steve.
Think for yourself! : an essay on cutting through the babble, the bias, and the hype /
Steve Hindes.
   p. cm.
   Includes bibliographical references.
   ISBN 1-55591-539-6
   1. Thought and thinking. I. Title.
   BF441.H524 2005
   160—dc22

                              2004030236

Printed in the United States of America
0 9 8 7 6 5 4 3 2 1

Editorial: Sam Scinta, Katie Raymond
Cover and interior design: Jack Lenzo

Fulcrum Publishing
16100 Table Mountain Parkway, Suite 300
Golden, Colorado 80403
(800) 992-2908 • (303) 277-1623
www.fulcrumbooks.com

This book is dedicated to the individual
who has a passionate curiosity for better answers,
the courage to think for himself or herself,
and the character to remain civil, studious,
and open minded for a lifetime.

Balance
evidence with uncertainty,
objectivity with passion,
lessons received with independent thought
Balance

Rule the World

First, you must have a natural curiosity.
I haven't, your quibbling all being so trivial, for I intend to rule the world.

Then you must prefer shades of imperfect evidence over the cherished myths of youth.
If a single word of them is false then I am lost; I shall never question them.

Then you must know how to find substantive information.
I haven't the time, for I must climb in authority and power.

Then you must know how to weigh the information, while ever seeking more.
I've neither the ghostly scale to weigh such vagaries, nor the patience—
I prefer to act, old man.

Then you shall indeed act most timely, while feeling fully the weight of your uncertainty.
Uncertainty is weakness in the eyes of the multitude, so it won't be found in
me. Uncertainty can be banished, or buried under pious prayer. There are
ways to ensure it shall never distract me.

Then you must assess your consequences, admit failure as well as success, and revise.
I shall be already to the next. My failures shall be only from the harassment
and traps of my enemies; but for those fools, my successes, which will be my
own, will have been all the greater.

Contents

Foreword

Thinking is skilled work. It is not true that we are naturally endowed with the ability to think clearly and logically—without learning how, or without practicing. People with untrained minds should no more expect to think clearly and logically than people who have never learned and never practiced can expect to find themselves good carpenters, golfers, bridge players, or pianists.

—Alfred Mander,
Logic for the Millions, 1947

For Whom Is This Book Written?

This book is designed for those who have been frustrated with the tendency of conversations to veer off track, to make questionable leaps in logic, to employ faulty data, to omit key considerations, or to be manipulated by bias rather than guided by enlightenment. The book gives the reader headlights and specific tools for navigating in the fog of complex contemporary issues. It provides an alarm system that can detect and alert the reader to the rhetorical manipulations and treacherous devices used to bamboozle him. It teaches techniques to weigh information and to identify or create a fair and valid argument. It also touches on critical lessons from history, science, religion, rhetoric, and mathematics that are not only helpful but critical in many aspects of life.

This book was written with the following persons especially in mind. Parents are afraid that as their children achieve independence and enter into the larger world, they could meet with and be confused by alluring strangers whose smooth talk draws them into cultlike beliefs contrary to years of home teaching. Some parents are afraid that their children will hold on too tightly to their upbringing and never develop the maturity and independence to improve upon it. This book gives people, young and old, a defense against the smooth talkers, while also keeping them open to better ideas.

It seems that high school teachers and university professors have taken a professional oath to despair that their students are not adequately informed or skilled at rational analysis. But why should students be skilled at rational analysis if no one ever taught them? Or if the professors themselves have weak skills? This book teaches these skills.

Persons in business have told me that when on the receiving end of a sales pitch, they may at times sense leaps of logic and manipulative language, yet it is difficult in the moment to identify the cause of their reservations. Only after considerable investment losses do they know the questions they wish they had asked. With this book, each side can enjoy a more honest and incisive exchange.

All normal, healthy personal relationships experience disagreement. Disagreeing is good because it reveals which facts, expectations, and viewpoints need to be clarified or reconciled. But not everyone knows how to disagree in a healthy way; not all discussions remain fair and logical. This book teaches the necessary skills of communication, especially when emotions are strong.

Lawmakers, and everyone who contemplates becoming a lawmaker—indeed, anyone who is eligible to vote—have a profound responsibility to acquire valid evidence, think objectively, and weigh issues without bias. Throughout history, civilization's greatest and darkest moments have often correlated to the degree of rationality practiced. Reason does not guarantee certain answers—indeed, it seems to ultimately expose the irreducible uncertainties. Reason does not guarantee any reduction in the variety of opinions or the passion with which they are held—again, it may well do the opposite. But reason does promise that the ideas explored will be of a broader perspective, a deeper fund of knowledge, and of a richer quality of evidence. A great civilization depends on great policy decisions and great policy decisions depend on reason.

Human nature itself needs the skills of reason. This is why so much of the content of this book has been a hot topic in many cultures throughout recorded history. Many of the ideas stated here were favorite topics of the ancient Greeks and of the enlightened caliphs in pre-Fundamentalist Islam. They are timeless lessons because they are immediately relevant, useful, and familiar to every individual who wants to make good decisions.

A Guide to Using This Text

Purpose

The purpose of this book is that it may help the reader, in some small way, leave the world a better place. Yep, I'm serious. Perhaps misunderstandings may be reduced, communications increased, problems solved. Where I scrutinize institutions or ideologies, it is not because I find them without merit, but because I believe we have a responsibility to build on their successes, improve on their weaknesses, reduce their irrationality, and hold them all to a higher standard of reason.

Usefulness

Every page of this book is designed to teach the reader how to pierce through the fallacies of the information he or she may encounter in boardrooms, schools, political debates, family disagreements, books, and the media. The reader will learn skills of detecting false logic and subtle manipulations of his or her opinion and of formulating a response that gracefully rises above such inadequacy or fraudulence by being more respectful, fair, and complete. Having learned to identify the fallacies of speech and thought, the reader may be surprised to realize how often such fallacies are used against him or her in everyday life.

Relevance

It is a great loss that some other books on logic are dry, buried on dusty shelves, because this is a topic that is anything but dull—it very often unleashes passion, delight, fury, and silent tectonic shifts of personal worldview. I hope that people will read these pages while considering the major issues of our day, regardless of their current position, regarding such topics as abortion, government regulation, gun control, environmental policy, religious conviction, immigration, campaign finance reform, taxation, health care policy, negotiation, and many others. These specific issues are not directly discussed in this book; we go farther upstream, to the heart of *why* people believe what they believe, *why* they vote how they vote, and *why* and *how* they make decisions in their daily lives.

Frankness

Many discussions of logic and rationality will veer toward topics that are emotionally charged with personal beliefs, from sex to established religion. I have tried to give such topics their due time and I have tried to treat the topics with a balance of sensitivity and frankness, while not shying away from important concepts just because they are sensitive. If, despite my best efforts, I have offended, then my consolation is that perhaps in pursuit of accuracy and logic I have accidentally offended everyone equally. I admire the reflection of Baruch Spinoza, seventeenth-century Dutch rationalist-philosopher, when he rejected an offer for a professorship at the University of Heidelberg so that his intellectual freedom would not be restricted: "I cannot be expected to teach philosophy without becoming a disturber of the peace."

Uncertainty

I will be the first to say that my observations and conclusions about reason might be entirely wrong. The reader shall, as always and as for all of us, have to think for herself. Furthermore, this book is not for people who are sure that they already have everything figured out. For them, there is no hope.

Thank you for taking a chance on this book; I think you'll be glad you did. Many people who read the manuscript have e-mailed me with favorite examples of fallacies and rhetoric from their everyday lives and from world newsmakers. I appreciate such examples and included many of them. Please don't hesitate to send me the ones you discover. Let me know the exact source, please, and if I can verify them, I may post them on my "Hall of Shame" Web site for most egregious uses of naked rhetoric. Send them to thinkforyourself@wispertel.net.

—Steve Hindes

Acknowledgments

I am indebted to many persons who have come before me who have also tried to explain reason and its proper application. As with any field of knowledge, this one grows with time, each person trying to climb "on the shoulders of giants"[1] to see a little farther, add a little more content, provide a little more clarity. I have climbed onto the shoulders of the giants listed below; whether I have added to the science is for the reader to decide.

I am particularly indebted to many of the ancient champions of rational thought, especially Plato and Aristotle. I am also indebted to many whose lives illustrated the concepts herein, especially Hippocrates, Socrates, Hypatia, Nicholas Copernicus, Johannes Kepler, Galileo Galilei, Giordano Bruno, Saint Augustine, Thomas Aquinas, and many other participants of contemporaneous and often conflicting schools of thought.

I have also learned a great deal from contemporary experts and scholars and am in their debt as well. I especially acknowledge Will and Ariel Durant, Carl Sagan, Richard Hofstadter, E. R. Dodds, Michael Shermer, and S. Morris Engel. I enthusiastically refer all interested parties to their important and influential works as mentioned in the bibliography.

Above all, I give my appreciation and love to my wife, whose patient support has made the production of this book possible.

The book is dedicated to my two sons: that they may think independently, choose wisely, take full responsibility for their decisions, live life to the fullest, teach the next generation to do the same, and, ultimately, leave the world a little better than they found it.

Introduction

What Is the Problem?

Imagine two middle-aged people bumping into each other at a store.

"Kristen! Wonderful to see an old friend here! Tear yourself away from those flat-screen TVs and tell me what you've been up to lately—how's life treating you?"

"Thomas, nice to see you too. When are you and Sally coming over for dinner anyway? Well, let's see. ... Not much really. The kids are keeping us moving all the time. Joe and I would like to make it to one more football game before the season is over. Work is going okay—never busy enough, it seems, until we're swamped, but getting by all right. You?"

Thomas smiles. "Sounds familiar. I guess the economy will pick up when we get a leader who knows what he's doing."

"I think our leader knows what he's doing, but it's the idiots around him that constantly stop him from accomplishing anything."

"Idiots?"

Kristen straightens up a bit. "He's finally in a position to reduce the tax burden on the producers in our society—finally let them get back to creating businesses, creating jobs, ... "

"Wait a minute—am I talking to Kristen?" Thomas smiles teasingly. "Since when do you believe that tax breaks for the rich are anything but blatant buying of votes from rich campaign contributors? This is new. ... I am so disappointed in you. I thought you were brighter than that. What cult radio show are you listening to?"

Kristen seems to hesitate. "Well, I had never read about the issue before. What irritates me are the whiners who want to bring the whole process to a stop—they want everything handed to them and want someone else to do the work. I tell you, it will be the destruction of our nation."

Thomas's measured tone borders on patronizing. "Don't you think that people living with yachts, multiple houses, and so many tax deductions that they hardly pay any tax at all could afford to pay a higher percentage of their income to give back to the country that made them so rich?"

"Who made who rich? These businesspeople make the country rich, not the other way around."

"What makes you think that?"

Kristen looks surprised at the question. "It's obvious. Everyone who knows anything about economics knows that. Don't you read Rush Limbaugh? Don't you read *The Wall Street Journal*?"

"Don't you read anything else? I don't read those because they are clearly biased. They are just mouthpieces for a particular audience who wants to be told that what they—the readers—already believe is absolutely true. They make big bucks stroking their readers and telling them that they are always right. They are the ones to whom the country owes its success. Anyone who disagrees is a parasite. Now, Michael Moore, on the other hand, there's a clear-thinking, evenhanded … "

"Maybe socialist rags such as *The New York Times* pander to a self-serving audience, but not the stuff I read and watch. And Michael Moore! He's a perfect example of Democrats—they hate America, hate everything our Founding Fathers stood for."

"Hold on now. The Democrats are the heirs of the Founding Fathers. The Founding Fathers knew something about the rights of the people, about "Blessed are the poor," about helping your fellow man."

"You can't be a true Christian and be a Democrat, given the party's position on abortion. … "

Yikes.

We'll leave these two old friends to the destruction of their relationship, to the mutual protection of their impenetrable close-mindedness, to the utter failure to learn anything from each other, and, ultimately,—and this is the main problem—we leave the meaningful issues of the day to continue to limp along without being addressed through evidence, logic, and reason.

Exchanges such as this happen at all levels of society, among ordinary citizens and at the highest levels of government. The encounters can easily veer into any of the other fascinating and critical issues of our day—immigration policy, health care costs, terrorism, the war in Iraq, taxpayer funding for art versus military defense versus care for the elderly versus subsidization of college education, and so on. The two characters above may have been family members at the dinner table wandering into a disagreement on the existence of God, the "proper" role of men and women in family, or how to handle the mother-in-law. In nearly every realm of our society, we encounter vital questions that beg for better answers, and in nearly every realm we do a very poor job of discussing the issues and discovering better answers.

The imagined dialogue above is loaded with very specific errors of thinking and errors of speech that poison our own thoughts and derail our discussions with others. These errors can be well-known, identified, and avoided, clearing the way for thoughtful and fair conversations that may actually produce new insights and better solutions—not to mention enriched and meaningful relationships among those with whom we discuss the issues, whether we agree with them or not. This book teaches very precisely what those errors are—how to detect them in the thoughts and speech of others and how to rise above them on our own.

Being able to think clearly and logically is a *learned* skill. What's more, each of us is responsible for *choosing* to develop these skills or choosing not to develop them.

Suppose you meet 100 people at a busy intersection in a big city and you tell them they have to find their way to a specific location fifteen blocks away. Surprisingly, they choose to blindfold themselves. Similarly, people who choose not to refine their ability to think clearly have also chosen to be blinded, and as they try to find their way through complex decisions, they tend to stumble into well-known pitfalls. This is especially true if they prefer preset notions that may not be based on valid evidence; that is, if they had been studying an inaccurate map.[1] These people may be good, decent, well-meaning, and hardworking, but misinformed and blindfolded as they are, when they go forth and attempt to find answers, they will often be spectacularly wrong, typically for these reasons:

- They choose to judge the merit of an idea by the degree to which it coincides with their current opinion, rather than by its inherent merit. If another opinion is different from their own, it is proof that the other opinion is flawed with incompetence or bias.
- They choose to judge the merit of an idea by the degree to which it serves their self-interest, personal sympathies, and physical, emotional, and spiritual comforts.

- They choose to see issues as black or white and have little patience for thoughtfully dissecting the options, making nuanced distinctions, or weighing with reason the relative validity of each issue.
- They choose to see only the strengths of their own opinions and only the weaknesses of alternative opinions.
- They choose to think that *wanting* something to be true is a sufficient reason to believe it is true.
- They choose to think they know enough on an issue to claim certainty.
- They choose to think that intuition, common sense, and personal experience are sufficient without balanced scholarship.
- They choose to believe it is always *other* people who do these things.

"We, the People," have chosen to govern by democracy, which means we will live by the opinions of the voters and of the persons we place in office by a constitutional process of our own design; thus, we had better have well-informed and reasonable opinions. We had better know what we are doing. Too often it seems that we do not. That we formulate and discuss opinions on the vital issues of our lives and our democracy so poorly is a very serious problem that threatens not just the quality of the lives of individual citizens but the very progress of our civilization.

The solution is far more complex than simple politeness because the problem does not start with how we *discuss* the issues; the problem starts with how we *learn* about the issues and how we *think* about them before we ever discuss them. The solution will require self-conscious and wrenching adjustments to cherished beliefs and daily lifestyles—and few people are interested in or willing to do that. Yet it is a grave responsibility of every citizen in a democracy. If we fail to live up to this responsibility, democracy itself will ultimately fail or be wrested away from us. What precisely anyone should believe on any political issue I will not say because I do not know. But I do know something about how people can navigate the depth and breadth of so many daunting problems and make up their own mind—and change it as they see fit—in a rational and informed way.

I. Complexity

Part of the problem is that the issues facing individuals and civilization are complex and numerous—people are naturally averse to grappling with problems that have many dimensions. The big issues of our day deserve well-substantiated, evidence-based analysis, yet people are often poorly equipped to study the complexities and to make evidence-based decisions.

Evidence must be retrieved from a person's fund of knowledge, and many persons' fund of knowledge is inadequate. The evidence and arguments that arise with any topic of discussion must be dissected with the skills of critical analysis, but those skills are vaguely defined, poorly developed, and haphazardly practiced. This book can give you the tools you need.

2. Evidence Doesn't Always Give Us the Answer We Want

Part of the problem is that we often do not want to bother with evidence— except when it suits us. We prefer our *feelings*. Truth is simply not much of a priority—it has value, to be sure, but many things seem to trump it, such as the satisfaction of seeming to have an immediate and certain answer, the comfort of myth, the pleasure of validation, or the spoils of power. Ultimately, most persons believe what they *want* to believe, not what the balance of hard evidence supports. I know that when I am trying to decide if I should claim an item in my family budget as tax-deductible, I decide that it is tax-deductible more often than the IRS does.

We believe what feels good to us, what serves our self-interest. Perhaps it feels good to believe that one's parents were right (or that they were wrong) or that one has the great fortune to be born in a time and place in which ultimate truth is embodied in the dominant culture and ideas. Perhaps what I believe gives me a sense of importance, awe, beauty, immortality, security, or any other pleasant sensation. We are capable of great intellectual contortions and selective blindness if doing so protects our cherished notions. It feels good to have an answer, to be certain, to not have doubt or a need to study, reverse oneself, or disagree fundamentally with the persons we love and respect. When many people are (justifiably) uncertain on what to do in complex matters, you can give yourself a great feeling by declaring you have it all figured out, know exactly what to do, and have an answer that transcends all moral ambiguities—even if you really don't. In general, preferred answers are very satisfying, they meet many needs, but truth is not necessarily one of them.

3. Rational People Live in a Saturating Culture of Irrationality

At times you may feel immersed in and swimming upstream against a culture of anti-intellectualism, vacuous entertainment, comforting self-delusions, well-intended misinformation, and near-fraudulent sales pitches, all if which may have a greater influence on individuals and society than scholarship and methodical introspection. The information we do receive is often passively acquired from self-interested "pushers"—those who have a disproportionate need for your concurrence—rather than through a combination of our own thoughtful search and the use of merit-oriented filters that would better nourish our minds. Even those who may be open to a healthy diet of even-handed, complete, and reliable information from many perspectives are instead force-fed mostly pith, spin, and myth in an indigestible swill. It is a diet that leaves us with the cognitive equivalent of rickets, scurvy, and beriberi—a mind that is ignorant one moment and fiercely certain the next, gullible with the fantastic and skeptical of the thoroughly substantiated. Such a mind is a shadow of its potential, and possibly irreparable. What can a reasonable person in such a culture do to sort through the empty filler, be better nourished, and find valid information with which to formulate evidence-based opinions in order ultimately to live a mental life that is full and vigorous?

4. We Start Adulthood Fully Loaded with Other People's Opinions

And we may end adulthood the same way if we never make a conscious effort to set aside cherished but unscrutinized ideas of youth and start over, examining those same ideas and every other idea ourselves from the per-spective of actual evidence, apart from the pressures of our immediate culture. To what extent do we actually self-consciously formulate our own opinion on any issue and to what extent do we passively absorb the opinions others gave us? Perhaps each of us is simply the graffiti wall on which others paint their opinions that we unthinkingly accept and reflect, our current opinion being the most recent sum of all previous paintings. We are born a blank slate, tabula rasa, and immediately afterward, our parents (typically) get first crack at us—a near monopoly, and the most time to paint a comprehensive set of opinions on our minds—long before we have any realization that this is happening. Other influences come along, seldom by our own choosing, and with varying degrees of heavy-handedness, frequency, repetition, and uniformity. We are left with an accretion that we resolutely declare our chosen opinion. Once we reach adulthood, perhaps we sense that the opinions of our youth were more often received rather than formulated and we reassess our opinions. Yet even then, we overwhelmingly confirm that, quite coinci-dentally, we still agree with the opinions of our youth, but now it is with even more certainty that our opinions are our own and have been carefully crafted after objective deliberation.

Ambrose Bierce, in his hilarious and irreverent work *The Devil's Dictionary* would have it thus:

Decide, v.i. To succumb to the preponderance of one set of influences over another set.

A leaf was riven from a tree,
"I mean to fall to earth," said he.

The west wind, rising, made him veer.
"Eastward," said he, "I now shall steer."

The east wind rose with greater force,
Said he: "'Twere wise to change my course."

With equal power they contend.
He said: "My judgment I suspend."

Down died the winds; the leaf, elate,
Cried: "I've decided to fall straight."

"First thoughts are best?" That's not the moral;
Just choose your own and we'll not quarrel.

Howe'er your choice may chance to fall,
you'll have no hand in it at all.[2]

People generally mean well, and although they are trying to influence us one way or the other, they are teaching us what they consider to be valuable facts and truths. We do well to listen, but to also scrutinize every idea for ourselves. Perhaps we should practice in our families what my medical school did for a class of mine. We were gathered in the lecture hall for the first hour of the first day of our first year of what we all knew would be a very long, thrilling, exhausting period of intense learning. The dean looked at us and said, "Half of everything we are going to teach you is wrong. We just don't know which half."

5. Confusing Intuition and Analysis
Part of the problem is that we need to better understand when to use explicit analysis more than the hunches of intuition and when the opposite would be more appropriate. Both of these types of thinking play a role in most human activities—riding a bike, balancing the checkbook, having sex, composing

music, and parking the car are just a few examples—but getting the right mix for any given task matters a lot. Thinking too hard when trying to have sex doesn't work very well, but thinking too little before you choose to have sex may be a disaster.

Analysis and intuition are not necessarily mutually exclusive, but they predominate at opposite ends of a spectrum:

Intuition	Tasks Using Both	Analysis
(cognitively fast and easy, but unreliable)		(cognitively slow and difficult, but accurate)

Pure analysis may be represented by complex mathematics, which few can do without profound attention to definitions, logic, and incremental process. Pure intuition—which is, by definition, free of any reference to data, experience, or external input—may be represented by what happens when a young child is placed in a room full of games that are familiar to the adults but unknown to the child: the unprejudiced child is likely to devise surprising, novel, innocent, and inappropriate ways to play with the games. This book is written for those who wish to tackle complex problems, which can be every bit as complicated as calculus, so the thinking should be heavily weighted toward the explicit analysis. However, all too often our legislators, businesspeople, and voters are "shooting from the hip," voting and enacting on issues by intuition—"gut feelings," and the most shallow of understanding.

In their defense, most of the time "intuition" is actually analysis on a subconscious level. For example, a professional ballplayer tracking a streaking ball and making a gymnastic catch may say his reaction was automatic and intuitional, but a great deal of conscious analysis ("Okay, it's coming this way") and subconscious thinking (tracking the flight of the ball and geometric extrapolation as to where to put the glove, based on innumerable dropped balls in Little League) had to occur with astonishing accuracy. Oh, if only the big questions in life allowed us as many chances and as many dropped balls as in Little League!

Why should we be so nervous about solving complex social issues by pure intuition? Why doesn't that work well? Two reasons: the first is that there is no data or experience on which to base the decision—neither personal experience nor information from others. For instance when a person is on skates for the first time or when an unprepared intern is handed a scalpel and pushed to the side of the operating room table for the first time, intuition will not produce the desired outcome. Having analogous experiences or input from a more informed colleague would be very helpful—but then it is not intuition, but rather explicit analysis, and that is exactly what we need more of.

Also, people may have a great deal of experience, but if that experience is in a system that never provided any consistent feedback from which to learn—those who "divine" for water with a forked stick, for example—then they are no better off the 100th time than when they tried it for the first time. They claim to function by instinct, and, despite their claims and their following, have been shown to be no more likely to find water than any individual picking spots at random. Why? Because, in fact, there is no "signal" or "connection" between water, stick, and diviner at all. There is no feedback at all, no accumulation of data or reference points for future attempts. Legislators who have not tracked the outcomes of previous attempts to solve problems are doing the same thing. Getting that feedback and considering it is explicit analysis and a very good idea.

The second reason that we should be skeptical of intuition as a reliable way to solve problems is that pure intuition, not based on any reliable experience, historically has a poor track record. People used to believe, intuitively, that vision was a result of rays of light coming out of the eyeballs and that the earth shook because the gods were trying to tell them something. And today, ask an untrained, inexperienced person objective questions (meaning questions that have specific, concrete answers, such as "What is the capital of Uzbekistan?"; "Which of these medicines will put the patient to sleep?"; or "What is the airspeed velocity of an unladen swallow?"[3]—and intuition will reveal itself to be just as poor a source of reliable answers as it ever was. Thus, pure intuition, if used at all, is better left to situations in which there is nothing to keep you accountable—no concrete data or experience for you to have to consider or no way to check your answer.

A *purely* analytic approach has its limitations too. First, it may be more rigid, cumbersome, and cerebral than what the situation calls for—persons who attempt to explicitly think through every move they make in sports or dance are likely to have an unsatisfying experience. Second, even for cases that do warrant maximum analysis, such as in the development of technology and discovery of natural laws, pure analysis is limited by the slowness of both the accumulation of valid data and that data's application. Third, once we have scrutinized our intuition to find out what the multiple fallible indicators are and have put them into an explicit formula, we might have missed some. For example, science may reveal that intuitive guesses about poverty and drug use contributing to crime levels are indeed valid, but that those two factors alone are poor predictors of crime levels because of the factors we intuitively suspect but for which we have no data—broken families, exercise of basic social manners, and so on. As a result, intuition may still outperform analysis. We have to go diving into our intuition again to find those other fallible indicators that our subconscious is using and that we had missed in our formulaic analysis the first time.

Perhaps the most strategic combination of these two types of thinking, intuitive versus analytic, is to approach problems very analytically and study them exhaustively from an evidence-based perspective. Only when the objective data runs out—and not when our interest, patience, or ego runs out—might we take the leap of intuition. That leap is an educated guess, a hypothesis, made with admitted uncertainty. Whether our leap proves successful or not will help us refine our analysis for the next time. But in every case, we want to climb as high as possible on the analytic knowledge before making the final exploratory leap. (I pause to make a distinction: this is no leap of faith. The leap of faith so often celebrated in some contexts is a proud act of conviction where there is, by definition, inadequate evidence to support the conviction. It is a leap indeed. I propose merely leaping to a hypothesis the way a frog leaps toward a lily pad—it may prove to hold and it may prove not to. But it is an educated guess, an experiment, and not a conclusion.)

Aristarchus of Samos (circa 310–230 B.C.E.), similar to Anaxagoras two centuries before him, was firmly convinced by observations and calculations that Earth was a sphere, not flat; that it rotated around an axis rather than the stars going around it; that the moon was much smaller than the Earth; and that the sun was much larger than the Earth; and he made ballpark calculations of relative distances from one to the next (all this sixteen centuries before Columbus's men sailed west, terrified of falling off the flat Earth!). But what he was unsure of was how all of this was put together as a whole. We do not know exactly how the idea of a sun-centered "solar system" came to him, given the lack of any knowledge of gravity as an astronomical force, but, as Charles Singer speculates:

> ... Aristarchus perceived that while the moon is smaller than the Earth, the Sun is enormously greater. This fundamental relationship may well have affected his thought, for it seems inherently improbable that an enormously large body would revolve round a relatively minute one.

It was his extensive collection and analysis of the available information first and then his educated intuition that led to his proposal of a sun-centered solar system and his modern recognition as the "Copernicus of Antiquity."

Thus, in the real world, we tend to make decisions with a combination of conscious analysis, subconscious analysis, and true intuition. We can choose between them—at any time we can abandon the gradually developing accuracy of analysis for the speed and ease of intuition. Also, we can learn from the success of intuition, study it, and put it to work in explicit analysis. Eventually, even for the most analytic minds, practical realities and necessity of a timely decision will require that they stand as high as they can on their mountain of analysis, fill in their remaining uncertainties with intuitive guesses, and take the hopeful leap. As any hang glider can tell you, your leap of intuition is more likely to give satisfactory results if you climb analytically to the highest ledge attainable from which to make your jump. Later, the results of that decision can be analyzed.

6. When Is "Common Sense" All-Too-Common Nonsense?

Common sense appears to be only another name for the thoughtlessness of the unthinking. It is made of the prejudices of childhood, the idiosyncrasies of individual character, and the opinion of the newspapers.

—W. Somerset Maugham,
A Writer's Notebook, 1949

Part of the problem is an exaggerated trust in "common sense." Common sense is really just another term for intuition, the weakness of which has already been described, but it is a term that is treated with reverence in a culture that celebrates black-and-white thinking, rapid decision making, and conformity. It contains in it an element of anti-intellectual pride: prosaic reasoning leading straight to a popular and certain answer feels very satisfying as a snub against the intellectuals who offer a long discourse on complexities that leads to an unpopular, guarded, and tentative answer.

However, common sense has a very mixed record for success and begs refinement. First, similar to intuition, it has frequently led us astray, producing many disastrous errors in history. For example, common sense was the driving force behind, in their days, witch burnings, economic exclusion of women, and slavery. People still cling to common sense to defend creationism (a clock suggests a clock maker, does it not?), tyranny (what is best for the people cannot be left to the ignorant masses), and other archaic and disproved notions. Second, frequently, common sense cannot predict or explain even that which we already know is real and true. For example, the way in which a common computer chip—the tiny human invention used frequently and daily by millions of common citizens—can produce text, images, and calculations completely defies common sense for all but the most analytic and informed computer fans and engineers. Third, common sense depends entirely on limited, circumstantial data, on common prejudices, on personal self-interest, and on whatever that particular era considers common knowledge—that the sun moves around the Earth, for example. As such it provides answers well in tune with one's own time and culture, but not necessarily at all in tune with ultimate reality or truth. The term "common sense" seems to put an air of respectability on a judgment process that may, in fact, be far too common and contain far too little sense. At best, it may be a source of hypotheses, but without further scrutiny, it is a very poor source of answers.

Almost all of us are born with some common sense in the same way that almost all of us are born with some hair: if we don't pay any attention to managing and improving it (which normally should start at an early age), we shall only embarrass ourselves, alienate most, and terrify some. Due to lack of timely reflection, we make our own lives surprisingly and unnecessarily difficult. Some of us will require more reflection and maintenance than others. Occasional devotees of unrefined common sense may argue that there is indeed some purpose to their determined self-neglect, that they derive an earthy satisfaction from their natural state. But we do not want to leave common sense, even more than hair and hygiene, in its natural, unimproved state. Such is the curse of the tabula rasa, the blank slate: each of us is born dumb as a chimp (no offense, but chimps do poorly, even in kindergarten), and only through a variety of learning experiences, which are often quite laborious if not painful—family, school, mistakes, friends, work, and so on—do we become educated.[4] On our tabula rasa, we each gradually inscribe and accumulate the knowledge that we will employ for all the decisions of life. The final fund of knowledge is highly variable between persons, in both quantity and quality. If we want to improve our common sense, we will have to improve our personal knowledge base on all issues and our skills of analysis and not simply rely on "common sense."

Egads, with common sense being so seriously flawed, how are we to approach the great questions of the day? "We, the People," hold the vitality or the failure of our democracy in our hands, but we are daunted by complexity, are not at all sure we want the truth, are plugged into fifty-two channels of *How to Marry a Millionaire*, and are just now realizing that our opinions were planted more than chosen and that intuition and "common sense" aren't going to cut it ... and we all have bad hair. What are we to do?

What we are to do, I passionately believe, is embodied in the following three principles that must be embraced by both society and individuals: we need a courageous *commitment to truth*, even when it is not what we wish it were; we need the maximum possible education for each individual, so that each of us may have in our own head a broad and deep *fund of knowledge*, directly and indirectly applicable to the main challenging issues of our day; and each person also needs, to the best of his or her ability, the *skills of critical analysis* for asking the right questions and sorting through the information with nuanced discrimination.

7. Irreducible Deficiencies

In researching many of the most interesting questions, we are faced with inadequate information, and despite our most laborious and sophisticated methods of obtaining better information, we will still be left with uncertainties. Yet often we have an obligation to attempt some sort of solution, even if it is admittedly experimental, and despite some success, we may well cause unavoidable errors and injustices. Reform of an education system, an energy policy, or a workplace rule may have to be tackled in the context of all these deficiencies. But we make decisions, and as we find better information, the decision will be reconsidered.[5]

Another inadequacy is the inherent arbitrariness of making distinctions within a spectrum—that is, of cutting a spectrum into separate groups. The groups may be truly different from one another in aggregate, but they are indistinguishable where they blur together; distinguishing the groups means drawing a line between the individuals who happen to be on either side of the line you have artificially created, and that is an unavoidable arbitrariness. For this there are many examples. Society may recognize that middle-aged adults can usually comprehend and digest unsettling images in movies of violence or sex (if they have a desire to), and society may agree that such would be bewildering and unhealthy for a six-year-old to see—but where do we draw the line separating the "ready" from the "not ready"? Often we choose the age of seventeen years (it is unclear to this author the basis for that decision). Is that to say that the day before a person's seventeenth birthday that he or she is emotionally too fragile but the next day is fully ready for the exposure? Is this to say that 100 percent of all fifteen-year-olds are not ready and that 100 percent of all nineteen-year-olds are ready? No. We admit to and accept the arbitrariness on the level of individuals very near the dividing line—in the micro level—to achieve useful and important distinctions between the separated groups generally—on the macro level.

Many times in which we must draw a sharp line through a grey area we are inevitably arbitrary to some degree. A government may calculate a "poverty line" so that families with incomes below it are offered assistance, such as taxpayer-funded health insurance or food stamps, and those above it are denied such assistance. Do we believe that there is a substantial difference between the families that make a single dollar less than the poverty line and those that make a single dollar more? No, but we establish criteria that are as reasonable and objective as possible for calculating where the line "should" fall and we use it as consistently and fairly as possible, until we devise a more reasonable and more objective way to recalculate it. Until then we cope with the "micro arbitrariness" for the greater good of the "macro rationality." If we have a naïve sense that any degree of arbitrariness invalidates the overall distinction and thus abandon the whole process of making any distinction whatsoever between the two extremes of the spectrum, then we have allowed ourselves to be paralyzed by the realities that confront us. Instead, we labor to find reasonable criteria, accept the irreducible arbitrariness, and keep an eye for a less arbitrary method.

Thus we are able to move forward with judgments that allow some students' exam scores to qualify them for further consideration for college admission despite being essentially indistinguishable from those who scored a few points less or that allow some law schools to make the listing of the "Top 100" despite being virtually identical to those that just barely did not. In every case, the placement of the dividing line shall remain admittedly debatable; all parties may fully admit the imperfections and injustices in the system, but perhaps the placement of the line should be revisited only when the alternatives seem less *arbitrary*, not simply *different*, from the existing standard.

Finding the Solution: Our Fundamental Guides

I. Commitment to Truth

Knowing the truth, good or bad, is a value in itself. No matter how elusive, exciting, daunting, disappointing, useful, complex, humbling, ugly, ambiguous, or inconvenient, intellectual honesty and integrity compel us to value truth for its own sake. Despite natural pain that comes with the loss of cherished myth, truth itself is both a goal and a starting point. We may delight at the beauty and complexity of the truth, but if that becomes our reason for seeking it, we risk shrinking from the pursuit when reality becomes disillusioning and uncomfortable. Nor do we revel too much in the shock and revolution that ugly truth may bring, for we may then abandon truth when it becomes mundane and "old-fashioned."

And yet, truth certainly does pay pragmatic dividends. The discovery of scientific principles has led to astonishing tools in medicine and the discovery of valid evidence serves justice in court. Truth and reality almost always serve better than ignorance or error, however blissful and comfortable the latter may be in the short term. The great purposes of life, as they are often regarded—understanding the reality of the universe we live in, comprehending life itself, being a "good" person (however you define that), and making a contribution to society, for example—are usually, if not always, served by being based on truth.[6]

2. Maximum Fund of Knowledge

One's fund of knowledge—the information carried about in one's mind—is developed through experience, including both formal education and experiences outside of formal education. Yet overall, the state of general education in the United States has been a disappointment—some would say a disgrace—since the founding of the country. While America may rightfully boast having many of the greatest universities and colleges in the world, and while there are indeed numerous specimens of dynamic and inspiring schools, the general landscape of public elementary and secondary schools is one of urban desolation, rural insularity, and suburban lassitude. Our great centers of higher education have done amazing work with the best and brightest from our high schools and with highly select students from throughout the world. Yet whether the schools fail to educate entirely or fail to educate beyond the scope of the standardized tests, the ultimate goal of producing informed and creative minds that can think for themselves, educate themselves, and improve themselves seems continually elusive.

The public debate on general education has recently centered on improving the elementary and secondary schools and on holding schools accountable to the "products" that graduate. I enthusiastically support that and would add to it that teachers in the general school system can't possibly turn out a consistently and highly refined product if they must start with remarkably poor material—grossly unprepared students—coming to them from homes and neighborhoods that do not emphasize the same skills and values of scholarship. We need not only schools that produce enlightened citizens, but a *culture* that produces enlightened citizens. If we leave the education of our students to the schools alone, the schools will never be able to compensate for what individuals, families, and society do not do.

One can imagine many strategies that could be used to teach young people to take personal responsibility for their own lifelong learning. For example, parents may consciously foster intelligent discussions within the home on meaningful issues of self and society by teaching the skill of detecting the ideas that require further research, modeling the specific skills of how to do that research comprehensively, reading together and discussing what was read, and helping the children articulate their discoveries and their responses in their writing and speech. Parents could do all of this, preferably in the context of a well-balanced life—intellect and athletics, science and art, seriousness and humor. … Parents may choose, for example, to stimulate the intellect for at least as much time per day as the children spends with passive entertainment, such as television, popular music, or cataleptic surfing of the Internet.

With a broad and deep fund of knowledge of moral tenets, politics, economics, history, science, literature, art, and all fields of study, we develop the basis for sophisticated decisions. Not from one of those sources, but from all of them. The pursuit of knowledge in these fields involves books, periodicals, musical instruments, teachers, museums, travel, and many other stimulating experiences, and thus is inevitably difficult and expensive. In fact, it is the work of a lifetime. What I wish to focus on and communicate here are the skills of critical analysis, with reference to the most immediately relevant lessons from history, science, dialectics, psychology, and mathematics.

3. Skills of Critical Analysis

Although it is common to make excuses for not trying to develop skills of critical analysis—and I am sympathetic to many of them; see chapter six, The Human Factor (page 167)—it is still nonetheless a choice for which everyone is responsible. Families, schools, lifestyles, and community institutions may teach the skills of critical analysis badly, or teach the converse of them; practicing the skills may be difficult and at times painful; and there may be few adult models to demonstrate scholarship and critical thinking. Yet, ultimately, each person has a responsibility for the degree of rationality with which he chooses to think because the clarity of thought we use can determine the state of the world and the advance of civilization. Not everyone would agree with that statement; just as many people do not believe that rationality, objectivity, truth, and concrete knowledge are critical factors in the advance of civilization. Further, many people do not believe in the value of "civilization," as it is commonly considered, or its advance at all. I hope that such skeptics (or true cynics?) would consider the concepts in this book and honor me with their feedback.[7]

For all those who want to refine their ability to critically analyze an argument for strengths and weaknesses or to make their own argument more fair, this book is offered as a guide. We will consider the lessons that are available to us on how to think more logically and rationally from several sources, including:

History: "Ontogeny recapitulates phylogeny." That is, the individual must retrace many of the developmental steps previously made by society itself. If we, as individuals and a society, are to rise more quickly and more wisely to the leading edge of cultural development, then we must consider some of the experiences, successes, and errors that humanity itself has made in the past.

Psychology: There are many easy errors of irrationality that can be avoided and many useful fundamental skills for processing thought that can be learned if we are aware of typical patterns of the mind.

Science: This field has an extraordinary track record for gradually untangling the mysteries of the real world; if we can learn how "evidence-based" thinking works in this field, then we may be able to apply "evidence-based" thinking in many other fields described above.

Superstition: Why is this such a powerful resource for answers? What are the advantages and disadvantages of employing this type of thinking in our most pressing questions?

Rhetoric: The art of persuasion can be used very effectively, whether to teach truth or error. Do you know when you are being manipulated and influenced by devices other than the fair presentation of facts?

Mathematics: The precision, accountability, and techniques of mathematics serve as both a model and a tool for our nonmathematical thinking.

Forward, then, into history!

Chapter One

A *Brief* History of Reason in Western Culture[1]

If we wish to learn how to think rationally, we must first understand the steps by which humanity learned to do so. It is a story that is amazing, tragic, inspiring, and ominous. Above all, it is instructive, for we are not at all above repeating the calamities of the past.

In the big scheme of the history of the universe, *human* history is astonishingly recent and brief. And within the frame of human history, the conscious effort to reason *well* is briefer still. In that very brief effort to reason well, the notion that *every individual, of whatever position in society, has a right and responsibility to reason well* is indeed so new that it seems to many a ridiculous idea.

All Truth passes through three stages:
First, it is ridiculed;
Second, it is violently opposed;
Third, it is generally considered self-evident.
> —*Arthur Schopenhauer (1788–1860),*
> *German philosopher and fun guy*

Newborns are fragile. There is no guarantee that reason, newly born among the masses and among our leaders, will automatically grow, strengthen, and prosper. Reason succeeds or does not because of conscious human effort; the effort must start all over again, from the beginning, with every new generation. Reason can be nourished and it can be neglected, it can be protected and advanced, and it can be lost. Indeed, reason has in the past been born into the world, survived infancy, grown to strength, and made vigorous and historic contributions to civilization—only to be attacked, actively suppressed, aggressively reduced, and, finally, voluntarily abandoned for the comforts of authoritarian absolutism. More than one great civilization has experienced such achievement and loss, as we shall see in the timeline of astonishing discovery and incalculable loss outlined below. In Western history, methodological rationality was quite nearly extinguished with the abandonment of Greek rationalism in the last few centuries B.C.E., at the cost of 2,000 more years of ignorance and savagery that we had been in the process of gloriously transcending. That collapse was one of the greatest failures in human history and it was a voluntary choice for which our culture is responsible. The scope and loss is beyond words; we must not make that error again.

For perspective on how recent the endeavor for humans to think clearly and logically is, let us consider first our place in the history of the universe. This was represented cleverly and effectively by the late astronomer and "popularizer" of science Carl Sagan.[2] It is his notion, with updated adjustments, that if we scale down the entire 13.7 billion-year history of the universe, from the big bang to the present moment, transposing it onto a single twelve-month calendar, then many of the landmark events would come at these dates and times:

One month=1.14 billion years
One day=37.5 million years
One minute=26,065 years

January 1, 00:00:01 A.M.: Bang. Matter begins condensing into perhaps 100 billion galaxies, plus other forms.

The scale of time in the cosmos is hard enough to grasp, but so too is the scale of size. Consider, for example, how far we are from our very next nearest sun, Alpha Centuri, and consider how long it would take to get there. It depends on technology that we are nowhere near developing. However, the fastest man-made objects to date are the two space probes, Helios I and II, developed by the United States and West Germany, which went 150,000 miles per hour en route to the sun (aerospaceweb.org). But how many miles is it to Alpha Centuri? If the star is 4.3 light-years X 364.25 days/year X 24 hours/day X 60 minutes/hour X 60 seconds/minute X 186,000 miles/second (the speed of light) = about 25.2 trillion miles, an object as fast as the fastest ever man-made object would cover the distance in = 25.2 trillion miles/150,000 miles per hour = ... = 19,217 years travel time to Alpha Centuri ... *one way* ... and again, Alpha Centuri is only the first closest sun of 100 billion in our galaxy, which is only one of 100 billion galaxies. And that's just looking up— using a microscope or a particle accelerator takes us into infinity in the other direction.

May: The Milky Way galaxy forms. It is composed of 100 billion stars, of which our sun is an ordinary one, quite peripheral from the brilliant center, midway in the dense galactic plane. Of these 100 billion suns, the very next nearest one, our next-door neighbor, Alpha Centuri, is 4.3 light-years away.

September 11: As with many of the stars, debris caught in orbit around our sun is forced by gravity into spherical aggregations. One of these will later be called Earth.

September 14 (4.5 billion years ago): Random chemical combinations happen to produce a symmetric molecule that, if chemically split, rebuilds from further random collisions. Both halves do this: the molecule has reproduced.

November 12: Microbes begin to exchange their DNA as a part of reproduction.

November 25: The first green plant begins to produce oxygen and nitrogen.

December 5: The Cambrian explosion—"life-size" life-forms appear.

December 16: Fish and vertebrates appear.

December 19: The plants move onto land.

December 21: Lungfish crawl onto land; insects and amphibians appear.

December 23: Trees and reptiles appear.

December 25: Dinosaurs live and die.

December 26: Mammals appear.

December 27: Birds appear.

December 30: The first apelike creatures appear.

December 31:
10:30 P.M. (a few million years ago): The first humans appear.

11:45 P.M. (7,000 B.C.E.): Fire is tamed.

11:59:17 P.M.: The domestication of plants and animals occurs.

11:59:32.4 P.M. (12,000 years ago): Agriculture and the first cities occur.

11:59:49.5 P.M. (5,000 years ago): Recorded history begins.

11:59:54.1 P.M.: Reason, having been used in everyday tasks to some degree in all human societies, crystallizes into a far more powerful form: the scientific method. This powerful organization of thought becomes a formally recognized approach for a tiny number of individual persons in Greece; they are the earliest people of whom we have a record of self-consciously pursuing the scientific method.[3] As of today, reason remains a new tool that has been used intermittently in public policy for only the last 2,600 years—the last 0.09 percent of human history.

Why this review of cosmological history? Because it is essential for the reader to understand how new and unfamiliar the very notion of systematic reason is; indeed, how new (and perhaps brief?) humanity itself is. People naturally tend to grant themselves, their species, and their convictions a special, and perhaps undeserved, place in the cosmos, so we are served well to remind ourselves of our very humble position, experience, and knowledge.

A more conventional timeline for the major events in the history of reason is as follows (many of the ancient dates are quite approximate). This is certainly a bare-bones rundown of some of the major events. The full story provides many examples of successes and errors surrounding the discovery of truth, including naïveté, grotesque murder, undeserved certainty, political ruthlessness, studied deception, courageous inquiry, catastrophic ideological warfare, forced suicide, high-minded balderdash, agonizing disillusionment, acceptance of uncertainty, and incremental progress. Many books recount these stories and personalities in vivid detail.[4] Rationality as a social force, freedom of thought and speech, and democracy seem to occur in correlation in history, so landmarks in the development of democracy are included in this timeline. Foremost is the constant tension between comfortable and authoritarian answers and the freedom and responsibility to think for oneself with the fullness of evidence. The reader is encouraged to look for patterns in this history and to consider if the forces that promote rationality and irrationality may still be very much present and in competition for the minds and the legislatures of society today.

624 B.C.E.: Thales of Miletus, in Asia Minor, was a Greek and the first known scientist. He believed in a mechanical universe that could be understood through careful observation and reason.

621 B.C.E.: During the renovation of the great temple in Jerusalem, the high priest Hilkiah discovered a scroll that commanded the Israelites to obey its laws and regulations, claiming that they were given by God and written by Moses. The scroll became what is now known as the central part of the book of Deuteronomy. Its actual authorship is disputed. Although the various books of the Old Testament were likely converted from many cross-cultural oral traditions to written forms from the tenth to the fourth century B.C.E., this is the first known reference in Israel to a sacred book.[5] Now stepping out of the fog of prehistory, the Israelite tradition later gave rise to three great world religions, and, for some followers, to a fundamentalist certainty about many crucial questions of society.

611 B.C.E.: Anaximander, also from Miletus, was an early astronomer and cartographer who attempted to explain the origins of the cosmos in material terms rather than through superstition. This willingness to set aside comforting and mythical beliefs for those supported only by objective assessment of evidence would be seldom appreciated or emulated then or now.

594 B.C.E.: Solon is elected archon, or chief magistrate, of Athens. His main legacy was to create a governmental system that guaranteed the supremacy of the rule of law: that every person, especially the rulers, had to obey the law as carefully recorded. It was the creation of this formal basis for how every person could participate in public life that was the necessary platform for the construction of democracy in Athens.

582 B.C.E.: Pythagoras of Sámos demystified and applied mathematics. He and his followers idealized numbers, believing they had a separate existence outside of the mind. This idea had a profound effect on Plato, who also looked beyond man's world to an ideal

world for perfection. Even mathematical equations were subject to mystical preference and certainty.

540 B.C.E.: Hecataeus, the first known skeptic of mysticism and religion, said, "The stories of the Greeks are in my opinion no less absurd than numerous."[6]

540 B.C.E.: Alcmaeon, also Greek, began the science of dissection. He described what we now call the eustachian tubes of the ear 2,200 years before the Italian anatomist Eustachio. On and off throughout history, the freedom to dissect the human body would be restricted as a violation of sacred ground.

510 B.C.E.: Cleisthenes replaced the tyrant Hippias and became leader of Athens. He solidified his position and reduced the influence of old aristocratic clans by appealing directly to the people and offering the population far greater power in political affairs. Democracy was born and developed as an established, complex, explicit system. While it included only the adult males, excluding women and foreigners, and depended on a slave economy, it was still a historical achievement of shared power.

479 B.C.E.: Once the Athenians, with substantial help from the Spartans, repelled the Persian invaders, Athenian democracy expanded into an empire.

470 B.C.E.: Democritus, born in Abdera, in Thrace, speculated that all matter is composed ultimately of atoms that are eternal and uncaused; their inherent motion is also eternal and uncaused. He proposed that nothing else is necessary to originate all things and events. This was a daring defiance of traditional notions that there is ultimately a divine cause for all phenomena. This atomic theory of matter was not pursued toward verification until 2,300 years later, in 1827, with the discovery of Brownian motion by the English botanist Robert Brown. Why do you think there would be such resistance to the notion of a mechanical structure of all things?

461–446 B.C.E.: Pericles ruled over The Golden Age of Athens and allowed more freedom of thought, speech, and inquiry than previously known. The rise of the Sophists (teachers) occurred, as well as friends and admirers of scientists such as …

468–428 B.C.E.: Anaxagoras gave mechanical explanations for celestial phenomena (for example, that the moon shines because of reflection of sunlight) and for biological phenomena (such as the origin of animals from natural elements). Such defiance of divine attributions for phenomena led to his persecution for impiety. The persecution may have actually been a political scheme to weaken his friend Pericles, leveraging the passions of the people on a less material but highly provocative issue to undercut Pericles's effectiveness in the very material but more prosaic issues.[7] The tactic of toppling a political opponent not through the greater merit of evidence and reason but through fanning the flames of a passionate and ignorant crowd would prove effective throughout history.

460–377 B.C.E.: Hippocrates of Cos, a physician, believed in a mechanical rather than mystic cause of disease and of all events. He taught careful observation and recording of actual phenomena, repeated verification of apparent cause and effect, and skepticism for fantastic and unlikely explanations. He was the first true scientist to leave substantial writing.

Men continue to believe in divine origin because they are at a loss to understand it. … Charlatan and quacks, having no treatment that would help, concealed and sheltered themselves behind superstition. … [8]

459–340 B.C.E.: Athens, the unique haven for free inquiry, overextended itself, neglected its domestic issues, and staggered with war against other Greek city-states, losing its empire and threatening all its rare and limited freedoms. Thucydides's account of it, *The History of the Peloponnesian War*, reflected his Hippocratic training—careful observation, reporting only the facts, and including no assumptions about divine purpose or interventions.

450 B.C.E.: Empedocles, a Greek philosopher and father of democracy in Agrigentum, speculated on the evolution of humans and animals from preexisting forms. Twenty-four centuries later, it may seem that no amount of evidence will ever be sufficient to dispel

cherished and mystical opinions on this question. Yet not all his hypotheses proved correct: he also speculated on the potential for transmutation of matter—that is, he contributed to alchemy, the pursuit of the "philosopher's stone," which would change base metals into gold. This proved to be one of many "scientific" delusions that would be not simply wrong, but a demonstration of how easily and frequently scientists from all ages can be very unscientific.

470–399 B.C.E.: Socrates, a native of Athens, lived and died applying the approach and skepticism of Hippocrates to many other fields and institutions, including religion and public policy. He insisted people question everything, especially the dictates of authority, and that people give better reasons for their own beliefs. He asked if mankind is capable of a natural ethic—a self-devised moral code for the benefit of society without the personal enticements and dire threats of a supernatural religion that places an all-seeing policeman and judge in the sky. He was condemned to death for impiety and for "corrupting the minds of the youth." His persecution and execution may also have been a political scheme brought on by Socrates's suspected preference for meritocracy over democracy. And his student ...

427–346 B.C.E.: Plato, also a native of Athens, advanced math and simplistic geometry, but slowed science generally by preferring the comfortable certainties of mysticism. He demanded that people direct their worries, aspirations, and service not on this world, but on another celestial realm, where everything was ideal. He cemented into Western culture the notion that individuals should ultimately live for the "next world" rather than for this one. He founded the Academy. And Plato's student ...

384–322 B.C.E.: Aristotle was a philosopher whose breadth of objective study and contributions to many fields remain an inspiration to modern scholars. His biology was far superior to that of Plato. He found analogous organs in animals that brought him near the theory of evolution, and he encouraged all ideas to be replaced as better, substantiated scientific ideas were developed. He may have hobbled scientific study of astronomy by approaching it as a mystic, despite using the observations of a lunar eclipse to argue the sphericity of Earth. He founded the Lyceum, the School of the Peripatetics, that began Stoicism—a philosophy of a mechanical universe, even to the point of everyone having an unavoidable fate. If there must be an Uncaused Cause, then Aristotle would have had it be God rather than the moving atoms of Democritus. And Aristotle's student ...

356–323 B.C.E.: Alexander the Great, a Macedonian himself, conquered vast territory from the Mediterranean to India, spreading Greek culture throughout, and died at age thirty-three. One of his many legacies was the foundation of the city of Alexandria, Egypt, which became the literary, scientific, and commercial center of the Hellenistic world, including being the site of the greatest library of the ancient world. With his death, Greece slipped back into civil war, and intellectuals fled from Athens to Alexandria. Philosophy began to de-emphasize natural science in favor of the study of morals (as if the two were in competition—are they?). Rome began its dominance of the Mediterranean and its neglect and erosion of the sciences.

440 B.C.E.: Philolaus proposed that Earth is not the center of the universe, but that it is one of many planets revolving around a "central fire" out of sight from the people on Earth.

430 B.C.E.: Hippocrates of Chios (and, much later, his student Euclid) developed geometry. The uncompromising logic, accountability to proof, and sheer fruitfulness had an enormous impact on the approach used in countless other fields. Students of geometry carried this taste for precise definition of terms, concise reasoning, and an unbroken chain of logic to many other fields, to the unending bedevilment of tyrants, priests, and parents.

342–270 B.C.E.: Epicurus of Samos founded the Epicurean philosophy that viewed the world as meaningless except for the pursuit of pleasure, including the pleasures of community service, education, and moderation. He did not value science. Why would it be unsettling that meaning and purpose of life be entirely a personal responsibility rather than being granted, ready-made, by a concerned omniscience?

300–200 B.C.E.: This era was the height of Alexandrian science and the Library of Alexandria. It was the largest intellectual institution in the world: nearly 1 million scrolls were housed here, representing the science, literature, and history of the known world. This was the center of the epic pursuit for the discovery and description of how the world really is, and not just how we wished it were.

300 B.C.E.: Herophilus began the practice of dissecting the human body publicly, despite traditional and religious taboos that regarded the human body as uniquely above explorations of its mechanics.

280 B.C.E.: Erasistratus described the nervous system, but this was lost until Bell rediscovered it, 1,900 years later.

280 B.C.E.: Aristarchus of Samos proposed that the Earth rotates around its own axis, producing the apparent movement of the stars, and that the Earth is just one of the many planets that revolve around the sun. He was charged with impiety.

287–212 B.C.E.: Archimedes, born in Syracuse, Sicily, and educated in Alexandria, Egypt, developed sophisticated mathematics and mechanics. His mechanical devices seem to us today to have been a potential seed for an industrial revolution. He was murdered by an impatient Roman soldier while doing a computation.

240 B.C.E.: In an elegantly simple yet clever study, Eratosthenes used the lengths of shadows in Alexandria and Syene to demonstrate the sphericity of Earth. By measuring the distance from Syene to Alexandria and then extrapolating, he calculated the circumference of the Earth, exceeding the correct figure by only 15 percent. He became the head of the Library of Alexandria.

200 B.C.E.: The Library of Alexandria began struggling under restrictions of thought and cultural apathy for science by Romans. In the timeless struggle between rationality and mysticism, the people were choosing mysticism again. The Age of Reason, unique to Greece and to all of history thus far, began to crumble and was not restored for forty-five generations of humanity.

Behind such immediate causes [for the Greek public to reject rationalism for mysticism] we perhaps suspect something deeper and less conscious: for a century or more the individual had been face to face with his own intellectual freedom, and now he turned tail and bolted from the horrid prospect—better the rigid determinism of their astrology than the terrifying burden of daily responsibility.

—E. R. Dodds, The Greeks and the Irrational

69 B.C.E.: Lucretius wrote *On the Nature of Things*, which argued "to so many evils religion has persuaded men." He proposed that we seek to understand the world instead through its mechanical workings, from its physically smallest units, *elementa*, to its natural overabundance of life. He attributed the diversity of life to the many spontaneous changes life-forms undergo, which then compete to survive and reproduce, resulting in gradual change over time. Eighteen hundred years before Locke and Hume, he proposed that reason cannot be the perfect test of truth since reason depends upon fallible sensation.

44 B.C.E.: Gaius Julius Caesar was assassinated. He had swept aside the 500-year-old republican form of government of Rome and become absolute ruler, setting off fifteen years of civil war.

27 B.C.E.: Gaius Octavius eliminated his rivals and succeeded Julius Caesar as supreme leader. The impotent senate declared him "Augustus," or "Holy One," believing, or seeming to believe, Octavius to be the Son of God, a common term and expectation of leaders of the time.[9]

4 B.C.E–A.D. 37: The life of Jesus of Nazareth. While historians are in general agreement about the existence and execution of Jesus, they have little agreement about his philosophy and teachings. Oldest surviving material reveals a wide diversity of equally certain versions of his message among, for example, the Proto-Christians, Marcionites, Gnostics, Ebionites, Theodotians, "Adoptionists," and Docetists, all of whom were certain that the others promoted a false, if comforting, version of Jesus' message. After his death, his story is embellished with miracles. No group was able to overwhelm the others until 367.[10]

A.D. 100: Hero discovered the principle of the steam engine. With cheap labor, including slavery, and a general apathy and common hostility toward science and innovation, the discovery was lost and unsurpassed until 1690, when Denis Papin "discovered" the steam engine.

A.D. 131–201: Galen, born in Pergamum in Asia Minor, advanced biology and anatomy. His work was scientifically adequate, but, in many ways, not as thorough or valid as previous scholars; his works survived because they accord with the political and religious expectations of the day. His work was not surpassed until Sir William Harvey in 1628.

A.D. 140: Claudius Ptolemy, an Alexandrian scientist, rejected Aristarchus in the *Almagest*, saying that the sun revolved around the Earth, and charged Aristarchus with impiety. This method of persuasion in academia lasts even until today, and its effect on astronomy in particular lasted for 1,400 years, until Copernicus vindicated Aristarchus.

A.D. 286: A weakening Roman Empire split administratively, controlled from Milan in the West and from Byzantium in the East.

A.D. 312: Constantine of Niš (in present-day Serbia), who had a history of divine visions, saw this time a cross of flame in the sun on the eve of the critical battle of the Mulvian bridge. He announced to his army that they faced a larger force and that they did so with the god of a small and marginalized cult, to which he had just converted, on their side. The army, accustomed to a similar god with a similar symbol, accepted this inspiration and went on to win the battle. Constantine consolidated his power throughout the empire and established toleration of Christians. He himself remained a pragmatist—"a statesman first and a Christian second."[11] Christianity grew from a persecuted Jewish sect to the dominant state religion, but, eventually, was divided in The Great Schism between Roman Catholicism of the West and the Greek Orthodox Church of Constantinople. The empire began a centuries-long process of transforming from a mosaic of regional religions to an empire that preferentially supported and often aggressively enforced one "correct" religion.

A.D. 325: The First Council of Nicaea created the Nicene Creed to settle the controversy about the Trinity being either a typical polytheistic artifact from other pagan traditions or sanctioned as consistent with monotheism.

A.D. 367: Of the many competing "Christianities," each with sacred writings, authorities and their followers finally wrestled one into dominance. Saint Athanasius, a bishop of Alexandria, sent to the churches under his jurisdiction his thirty-ninth festal letter. This letter, preserved in a collection of annual Lenten messages given by Athanasius, listed as canonical the twenty-seven books that remain the contents of the New Testament. All other versions of Christianity and their writings were declared heretical and became "apocryphal." Augustine of Hippo pushed the letter's acceptance at the Synod of Hippo in 393, and this largely settled the issue of which writings would speak for Christianity. Subsequent debates on the message of Jesus would amount to little variation from this core orthodoxy, relative to the vast differences debated in the first three and a half centuries after Jesus.

A.D. 410: The Goths sack Rome. It is the first time that this has happened to the ancient city since the Gauls in 390 B.C.E., which was eight centuries earlier.

A.D. 412: Cyril was elected patriarch of Alexandria and launched a zealous and merciless persecution of non-Christian people and activities. Through politics, intimidation, and direct action, he shut down churches of Christian sects other than his own. In retaliation for Jewish attacks on Christians, he instigated assaults on the Jews of Alexandria, destroying their homes, and finally driving them from the city. The Church later declared Cyril a saint.[12]

A.D. 415: In one of the riots during the rule of Cyril, Hypatia, a female mathematician, philosopher, and head of the Library of Alexandria, was stripped, skinned alive, and her remains burned by Christians offended by the intellectual pursuits of the library.

Within a year, they burned the Library of Alexandria to the ground. There is no evidence for the supposition that Cyril was instrumental in her death.

A.D. 426: Augustine, of Tagaste, Numidia (now Souk-Ahras, Algeria) wrote *The City of God*. In this, as well as in other works, he opposed many contrary Christian sects, especially Manichaeans and Donatists, and established the notion of Christianity as having divine right to rule all nations. His influence on Christian doctrine was broad and mixed, promoting, for example, profound misogyny and an uncomfortable toleration of sex, even in marriage, if at all. The church later made Augustine a saint.

A.D. 476: What was left of Rome fell to the tribes of the central Europeans.

Circa A.D. 550: Justinian I closed the schools of Athens, finding their intellectual freedom and ideas a threat to state order.

Circa A.D. 570–632: Muhammad of Mecca, a trader, experienced a vision in which the archangel Gabrielle whispered into his ear the teachings that would soon be written down as the Koran. Muhammad advocated social reforms contrary to preferences of some authorities, which gained for him a following, but also pro-pelled him into life as a military leader. Victories made him the most powerful man in Arabia. After his death, his followers embel-lished the story of his life with miracles. His legacy included the equation of religion and politics.

A.D. 630–733: Islam spreads explosively through conversion and military conquest, establishing itself from India to Spain. It would dominate the Mediterranean for 500 years. Moslem intellectuals, who were not yet seen as an affront to Islam, seized and studied the available texts of Greek antiquity, but the promotion, use, and pro-tection of them was very limited. While the majority of scientists and philosophers of the Moslem world were Persian, Turkish, or Berber, Arabs also participated in this surge of intellectual inquiry of Islam.[13]

A.D. 780–850: Muhammad ibn-Musa, perhaps the greatest mathematician of medieval times, greatly developed the science of algebra and introduced the name itself to the West. His *Calculation of Integration and Equation* gave solutions of quadratic equations and was used as a principle text in European universities until the sixteenth century. He also developed astronomy, trigonometry, and geography.[14]

A.D. 800–1000: Abbas, an uncle of the prophet Muhammad, established a dynasty of caliphs that ruled practically for a short time and symbolically for 500 years. During the ninth and tenth centuries, a great effort was made to acquire and develop the knowledge of medicine, mathematics, and other sciences from earlier and surrounding cultures. Abbasid al-Mamun founded the so-called House of Wisdom (Dar al-Hikma) in Baghdad (a city founded by this same family). Reminiscent of the Library of Alexandria, this institution was an enormously productive center for the acquisition and translation of written works—in this case, mostly of science and philosophy from the West, particularly from such authors as Aristotle, Plato, Euclid, Galen, and others,[15] but also of the sciences of the Hindus. Original research was devised and conducted to advance beyond those discoveries. A scientific academy, observatory, and public library were also established there. "These astronomers proceeded on completely scientific principles: they accepted nothing as true which was not confirmed by experience or experiment."[16] The resulting synergy of culture and ideas brought Islamic academic creativity to its greatest height and far surpassed the achievements occurring in the West at the time. It was through the preservation and advancement of knowledge here that the West eventually rediscovered the freedom of independent thought and the responsibility to question assumptions.[17]

A.D. 873: During this time, the numeral zero was used for the first known time. Although it was found in an Arabic document, it was probably adopted from earlier use in India; original sources were lost. This development came thousands of years after many cultures had developed numeral systems otherwise. The development of zero was an astonishing advance in human reason as it represented a willingness to not assign a value, or that the value is "no value."

However, its logical equivalent, the willingness to answer a question with "I don't know," remained rarely employed, as persons still felt compelled to make up answers when they did not know, thereby throwing off the appropriate "arithmetic" of logic.

A.D. 973–1048: Abual-Rayhan Muhammad ibn Ahmad al Biruni ("al-Biruni"), an Arab of present-day Uzbekistan, made original advances in many fields of science and literature and became a favorite in Mahmud of Ghaznis's court. He hungrily studied religions and cultures, with unusual impartiality. In the preface to his *Vestiges of the Past*, he wrote, "We must clear our minds from all causes that blind people to the truth—old custom, party spirit, personal rivalry or passion, the desire for influence."[18]

A.D. 980–1037: Avicenna (Arabic, Abu Ali al-Husayn ibn Abd Allah ibn Sina), Iranian Islamic philosopher and scientist of medicine, astronomy, and logic, was born near Bukhoro (now in Uzbekistan). His textbook, *The Canon of Medicine*, was still used in the universities of Montpelier and Louvain until 1650. Avicenna was the last of the great philosophers of the Moslem East. He himself was a target of orthodox pressure for being willing to disagree with the teachings of the religion in which he was raised. He denied personal immortality, God's interest in individuals, and the creation of the world in time. More generally, his philosophy of inquiry and science became identified with other Shi'ite heresies and unconventional mysticism; Avicenna became a primary target of the orthodox backlash led by al-Ghazali, whose fundamentalist text, *The Revival of Religious Sciences*, is today considered by many Moslems second only to the Koran. Despite these pressures, Avicenna spent the last fourteen years of his life as scientific adviser and physician to the ruler of Eşfahân (Isfahan).[19]

A.D. 1095: Two hundred years of Crusades began, purportedly to retake Jerusalem from the Moslems, but also in pursuit of many other interests—commercial, political, military, and religious. The extent of expansion of Islam into Europe was reduced.

A.D. 1106: Abu Bekr ibn Bajja ("Avempace"), a Spanish-born Moslem philosopher, may have been one of the early victims of rising anti-intellectualism in Islam. Yet another accomplished polymath, he was forced from his residence with the Moslem governor by encroaching Christians, but arrived among traditional Moslems who accused him of atheism. He died at age thirty, allegedly by poison.

A.D. 1107–1185: Abu Bekr ibn Tufail ("Abubacer") wrote a romantic parable in which the hero, a supremely enlightened autodidact, tried to bring his universal understanding to the common people of the world and was rejected. He concluded that people cannot be trusted to conduct themselves in a socially stable and productive manner and so must be compelled to do so through religion, which provides the lure of rewards, threats of punishments, escapes of fantasy, and reassurances of afterlife as essential incentives.

A.D. 1150: Ibn Habib was put to death for studying philosophy.

A.D. 1195: Averroës, of Córdoba, Spain, an Islamic philosopher, jurist, and physician, was exiled for his view that reason takes precedence over religion. He also studied theology, philosophy, and mathematics under the Arab philosopher Ibn Tufayl and medicine under the Arab physician Avenzoar. He was restored to favor shortly before his death.[20]

A.D. 1200: This is an approximate date at which the Moslem intellectual experiment was overwhelmed by a greater preference for mysticism. As with the Hellenes, the people of Islam fled from the freedom to think independently. They found an incompatibility, not previously recognized or enforced, between their fundamentalist faith and intellectualism. Book burnings become commonplace. As with the West's descent into orthodoxy and mysticism 1,500 years prior, Islam followed a similar path, just barely handing off to the West the works that gave rebirth to its classical and intellectual culture. Although the seeds were planted in the West, there was little growth for more than 300 years.[21]

A.D. 1210: The Church Council of Parish forbade reading much of Aristotle.

A.D. 1215: The Magna Carta, in England, established that the king is not above the law. This was the first defeat in the dominant practice of the divine right of kings. It was the first time since the fall of the Republic of Rome that anyone but the king was acknowledged to have rights and liberties—the circle of subjects included as those deserving of these rights and liberties inexorably expanded over the centuries. Specifically, it established the need for taxes to have the consent of the taxed; the custom of summoning to the royal council not just barons but elected representatives of towns and counties; and the convenience of dealing with petitions at enlarged meetings of the king's council.

A.D. 1225–1274: Saint Thomas Aquinas, Prince of the Scholastics, established Scholasticism—the attempt to reconcile faith and reason, so long as conclusions always concur with established doctrine. It dominated intellectual inquiry for 400 years.

A.D. 1231: Pope Gregory IX formally created the Inquisition to judge those accused of heresy and to severely penalize those found guilty. It was the latest and most formal manifestation of a very old effort.

A.D. 1252: Pope Innocent IV sanctioned the use of torture in the interrogation of suspected heretics.

A.D. 1255: The works of Aristotle became required reading at the University of Paris.

A.D. 1258: The barons of King Henry III, the Oxford Parliament, forced Henry to accept rule by a baronial committee. This was the first significant expansion of rights after the actual Magna Carta.

A.D. 1295: The so-called Model Parliament of Edward I contained, for the first time, all the elements of a mature parliament: bishops and abbots, peers, two knights from each shire, and two representatives from each town.

A.D. 1376: In the fourteenth century, England's Parliament further evolved and gained power. It split into two houses (the House of Lords and the House of Commons), gained control over statutes and taxation, and created impeachment (1376).

Circa A.D. 1400: The ancient classics of Greece and Rome, previously known to very few scholars, became a sensation among the literate classes. With the Turks encircling Constantinople, the city from which so many classics were being unearthed, a desperate attempt began to find and rescue any manuscripts before another sacking of this historically beleaguered city. Those who read and contemplated the significance of those works were so strongly affected that they manifested the ideas in the dramatic invigoration of the Renaissance, the Reformation, and the Enlightenment. The Catholic Church, already Roman in structure, dress, and conduct, financed architecture, paintings, sculptures, and a carefully limited enthusiasm for literature, much based on the discoveries of the ancients.

A.D. 1415: John Huss was burned at the stake for objecting to abuses by the Church. The Hussite Wars erupted between Catholics and proto-Protestants and ravaged Bohemia for seventeen years.

A.D. 1434: Cosimo de' Medici the Elder, a shrewd politician, established political dominance in Florence. His descendants ruled Florence for almost three centuries; political and financial ties made them indispensable to the Church. At the same time, their independence of thought cultivated a return to the ancient architecture, appreciation of the beauty of the human form, and culture that contributed dramatically to the Renaissance, and thus to the Reformation, and then to the Enlightenment.

A.D. 1453: Constantinople fell to the Ottoman Turks. It remained in Moslem control until 1922, when Mustafa Kemal Atatürk made secular and Western reforms.

Circa A.D. 1470: The newly invented printing press was reproduced throughout Europe and both Bibles and ideas contrary to scripture became increasingly available. One of the favorite topics was the abuses and absurdities of the Church.

A.D. 1510: Raphael painted the School of Athens, enshrining the timeless debate as to whether our ultimate concern and focus should be the next world or this one: the elder Plato gestures toward the heavens, while Aristotle gestures toward the real world right here on Earth.

A.D. 1511: Desiderius Erasmus published *Praise of Folly*. Though a priest, he ruthlessly satirized the Church, using reason and rhetoric to call for reform of Church practice and theology.

A.D. 1517: Martin Luther declared that the accumulated structure and policies of the Church, such as sacraments, rituals, and art—especially those that seem to have been absorbed from pagan influences—were deleterious to the relationship with God. The only ultimate authority that he recognized was the Bible and the individual's conscience. This freedom of individual thought escaped Luther's control and contributed to a widespread and unintended fragmentation of the Church and to open questioning of even the fundamental assumptions about the Bible and Christ still advocated by Luther.

A.D. 1524: The Peasant's War between Catholics and Protestants occurred. During this war, the Protestant Anabaptists were persecuted by the Lutherans, other Protestants, and the Catholics. The savagery of the Protestants caused Luther to speak out against his own supporters in *Against the Murdering, Thieving Hordes*.

A.D. 1530: Paracelsus, Philippus Aureolus, a pseudonym of Theophrastus Bombastus von Hohenheim (circa 1493–1541), contributed healthy skepticism to medicine, especially in his questioning that diseases originate internally by an imbalance of "humors." Instead, he insisted that diseases were often the result of external forces entering the body. At the same time, he was an early believer in the therapeutic benefits of magnetism. This cherished belief persists to this day, despite there being no credible scientific evidence.

A.D. 1534: The Anglican Church was established by Henry VIII, whose middle ground resulted in execution of those who denied the main tenets of the Catholic Church while also persecuting those who refused to recognize the ecclesiastical supremacy of Henry over the Pope.

A.D. 1542: The Holy Roman and Universal Inquisition was created by Pope Paul III to stem the spread of Reformation doctrines and also to specifically oppose the unorthodox writings of theologians or high churchmen. Until the start of any given trial, the accused were not allowed to know the charges put against them or the evidence against them. The accused were not allowed to have counsel; every trial was heard by a council of ten church cardinals; and the testimony of two witnesses was considered proof of guilt.

A.D. 1543: Nicholas Copernicus, himself a monk, physician, and astronomer, published *On the Revolutions of the Celestial Orbs*, which outlined the data and calculations of the Earth's place, not in the center of the solar system, but as the third planet from the sun. Under the threat of excommunication, the Church forbad that it be read. Eventually, the Church allowed the idea to be discussed "if only as a theory, not as a proven fact." The Church removed the threat of excommunication for this book in 1828—285 years later.

A.D. 1546: The Schmalkaldic War between Catholics and German Protestants ended after nine years with the Peace of Augsburg. Lutheranism was officially acknowledged, if not morally accepted, and the supremacy of the pope was destroyed.

A.D. 1559: The *Index of Forbidden Books* was first published by the Catholic Church. Any person who possessed, read, sold, or transmitted a book on the list without ecclesiastical dispensation was excommunicated. Last published in 1948; discontinued use in 1966.

Circa A.D. 1560: Prospero Alpini of Padua discovered the doctrine of sexual reproduction of plants—1,800 years after this had already been done by Theophrastus (372–287 B.C.E.).

A.D. 1563–1592: Michel de Montaigne did for freethinking in France what Bacon (page 39) was doing in England—popularized skepticism, reason, and research; advocated frequent admission of uncertainty; and made enough acquiescence to the Church to find himself involuntarily forced to be mayor of Bordeaux by Henry III.

Circa A.D. 1566: Tycho Brahe, a Dane, began his reclusive and massive collection of measurements of the stars. He died before he could understand the significance of his data.

A.D. 1572: Saint Bartholomew's Day Massacre—Catholic mobs murdered thousands of Protestants in France. The number murdered cannot be determined accurately but is estimated to be 2,000 to 100,000.[22]

A.D. 1575: The great observatory of the Moslem world, built in Istanbul by Taqī-al-Dīn, was ordered destroyed by the chief mufti. His elite corps of soldiers, the Janissaries, swiftly fulfilled the orders.[23]

A.D. 1600: Giordano Bruno, an Italian monk, philosopher, and poet, was found guilty of supporting the Copernican model of the solar system, for reasoning against the Church dogma on the Incarnation and the Trinity, and for other heresies. He was burned at the stake in Rome by the Inquisition.

A.D. 1603–1620: Francis Bacon appealed to James I, king of England, to subsidize a universal effort in all sciences and the development of reason. The Great Renewal began, and in many fields of science, the water finally rose above the mark left by Aristotle and others 2,000 years before.

A.D. 1604–1619: Brahe's assistant, Johannes Kepler, a German astronomer, found that Brahe's data did not correspond to his idealized belief of the structure of the universe. He relinquished the mysticism of his lifetime to accept the actual calculations: he revealed the elliptical movement of the planets, clarified the Copernican system, and set the stage for Newton's discovery of gravitational laws. That the solar system moves so mechanically and predictably, with no particular aesthetic or special role for man or Earth, was a staggering blow to the credibility of the Church, which has based much of its authority on the certainty of such claims.

A.D. 1608: Hans Lippershey, Dutch spectacle-maker, described his invention of the telescope. Galileo Galilei, Italian physicist and

astronomer, quickly improved and patented it, and retired to a life
of comfortable personal research.

A.D. 1618–1648: Christians again slaughtered Christians over religious certainties in the Thirty Years War. Involving most of the
countries of Western Europe, the war achieved mortality rates to
rival the plague. Twenty percent of the entire population of present-
day Germany died, and as much as 50 percent in other regions in
Europe. It was "A prolonged and extremely cruel war." The war's
end established that the Protestant Church had as equal claim to
the Christian Church as the Catholic Church.

A.D. 1619: Lucilio Vanini was burned at the stake for proposing that
humans evolved from apes.

A.D. 1628: Sir William Harvey demonstrated the circulation of the
blood, and, for the first time, anatomy moved beyond Galen of
1,500 years before. The notion that the heart functioned as the seat
of the soul came into question; man himself was increasingly seen
as a fundamentally mechanical phenomenon.

A.D. 1633: Galileo's longtime insistence that the Copernican system
is clearly fact and his demonstration of the existence of vast stars
previously unseen led to his arrest by the Inquisition (June 1633).
Galileo, after being shown the instruments of torture, recants, but
was kept forcibly under house arrest until his death nine years later.

Circa A.D. 1640: René Descartes, French philosopher, mathematician, and scientist, popularized that all the universe is mechanical,
following natural law only—except for the human mind, which, he
concluded, is the window to and proof of God. He shifted all of
European philosophy to epistemology—understanding what it is
to think, what is it to know. "Why should we trust ideas? How can
we be sure of the validity of anything we think we know?" Unable
to answer, he concluded the answer is God. Yet the Calvinist
Protestants still felt he threatened their view of predestinarianism
with his conception of a mechanical universe. Threats from the
Church led him to destroy almost all his later work; public discussion of his ideas was forbidden.

A.D. 1648: After ninety years of warfare in the Netherlands, the Catholics retreated under the terms of the Peace of Westphalia.

A.D. 1649: Archbishop James Usher, Irish prelate and biblical scholar, declared that on October 22, 4004 B.C.E., at exactly 6:00 P.M., God created the world. His calculations were based on the presumed life spans of the persons mentioned in the Bible.

A.D. 1652–1698: French Protestants, the Huguenots, were locked in a generation of warfare with French Catholics.

A.D. 1687: Isaac Newton, English physicist, scientist, and natural philosopher, published *Philosophiae Naturalis Principia Mathematica* (literally, *Natural Science through Mathematical Principles*), or *Principia*. With devastating coherence, it seemed to reduce the sacred universe to faceless natural laws. Gravitation, inertia, and the laws of motion were laid bare through the mathematical science that he and Leibniz invented in the process: calculus. Only the less educated or most faithful continued to insist that the scientists were wrong about this issue. The rationale of the steadfastly devout was that the scientific ideas seemed to undermine the credibility of the Church, the idea of the existence of God as previously conceived, and of humanity as the special creation in an otherwise mechanical universe.

A.D. 1688–1689: England's Glorious Revolution took place, which established parliamentary sovereignty over the king.

A.D. 1689: The Toleration Act of England extended toleration to all Christian sects except Catholicism. It also excluded all other doctrines, including Judaism, Islam, deism, atheism, and paganism. However, it was the first national legislation of any degree of religious toleration since the ancient Greeks.

A.D. 1690: Locke published *Essay Concerning Human Understanding*, in which he proposed that the human mind is born a blank slate, with no inherent ideas, and experience supplies all information and ideas.

Denis Papin "discovered" the principle of the steam engine—again.

A.D. 1733: Parish priest Meslier died, leaving on his deathbed *My Testament*, a scathing indictment of the Church and its dogma.

A.D. 1739: David Hume, Scottish historian and philosopher, published *A Treatise of Human Nature*, claiming that our lofty and autonomous reason is neither at all, but a machination of ideas with inevitable output. Nonetheless, he used his own machinations to demolish the notion of miracles and other superstitions. Eighteen hundred years after Lucretius and 120 years before Darwin, he proposed that organisms' designs are (in Durant's words) "not from divine guidance but from nature's slow and bungling experiments through thousands of years."

A.D. 1760s: Voltaire led a frontal assault on irrationality and superstition.

A.D. 1771: *Encyclopedie* was first published. With wit and covert placement of passages, science and opinion that was contrary to the preferred doctrine of political and religious authorities was made public. It was edited by the French philosopher Denis Diderot in Paris. Contributors were many great French writers of the day, including Montesquieu, Voltaire, and Jean Jacques Rousseau, and the German Friedrich Melchior, Baron von Grimm.

A.D. 1775: Franz Friedrich Anton Mesmer (1734–1815), an Austrian physician, used a combination of magnets and hypnotism as medical therapies and as a way to fame. Although this particular bit of pseudoscientific delusion was debunked even in his own day, the demand for answers and solutions through "junk science" was readily met by many other eager suppliers, even to present-day.[24]

A.D. 1776: Despite subsequent celebrations, this was hardly a time of independence and national birth. The Declaration of Independence was, in effect, a declaration of war, as the signers knew that would be the effect. Even if war could be won, which it barely was, and then only with critical assistance from France, the new government had to survive its first very ineffectual Articles of Confederation. Eventually, those articles were abandoned to attempt a frightening return to Federalism as embodied in the Constitution.

A.D. 1785: Scottish geologist James Hutton developed modern theories of geology, including calculations of the millions of years necessary to produce Earth's current landscape. This theory set him at odds with the Church, which held that the Earth was less than 6,000 years old. Charles Lyell, in *Principles of Geology*, 1830, courageously attacked the theological prejudices and the use of scripture as the ultimate reference for geology.

A.D. 1786: Thomas Jefferson's Act for Establishing Religious Freedom was passed in Virginia, assuring liberty for all types of religious belief and nonbelief. The act was passed while another bill, proposed by Patrick Henry, to tax all citizens for the support of Christian clergy, was rejected, largely due to the eloquent argument against it by James Madison in his "Memorial and Remonstrance Against Religious Assessments." Jefferson's act would become a model for the secular passages of the U.S. Constitution.

June 21, A.D. 1788: The U.S. Constitution went into effect. In many ways, it was at its inception the summation of political lessons from the centuries, including a prudent fear of irrationality and of inadequate policy assessment.

At the Constitutional Convention in Philadelphia, the authors agreed there was sufficient evidence in history to legitimately fear too much power and too little consideration in the hands of either a strong executive or, at the other extreme, in the hands of the common people.

To avoid their first fear, the original proposal for the new government, authored by James Madison and modeled after England's parliamentary system, was for a government dividing federal power and policy assessment among three branches: executive, judicial, and legislative. A system of checks and balances was established so that no one branch would dictate to the people. For example, the legislative branch was responsible for passing laws, but the executive branch could veto legislation; the veto could, however, be overridden if two-thirds of the members of both houses of Congress voted to override. Even then, the Supreme Court could find the legislation unconstitutional and thus void. The people could reverse the Supreme Court by voting for a constitutional amendment. The president appointed the members of the judicial branch with the advice and consent of the Senate. Once appointed, the judges served independently of either the executive or legislative branches. The judges, as well as the executive, could be impeached and dismissed for high crimes and misdemeanors.

And since the authors believed that the common people were ignorant, easily misled, and were not always able to select the best person to represent them, they avoided their second fear by writing into the Constitution a series of devices to limit the ability of the common person to directly affect the government. Foremost, the electorate would be limited to white males over age twenty-one who met various additional standards in each state, such as poll taxes. The legislature would have two houses, with the lower house (the House of Representatives) being elected directly by the people and the upper house (the Senate) being selected by the legislature of each state rather than by the people directly. The president, who was given extensive appointive and military powers, was to be selected by an electoral college made up of people appointed from each state's legislature rather than by the people directly.

The very creation of a constitution that granted unprecedented yet still limited power of the people and that shifted much power from the states to the federal government was effectively a coup d'état—it was purposely done behind closed doors so as to keep the public unaware and it ignored the existing Articles of Confederation's mechanism for change. Had the people known, or had the established mechanism been used, the Constitution likely would not have passed.

Another lesson from history that the mostly Christian Founding Fathers heeded was to never pretend that this law was from God and infallible, but that it was entirely a human construction, unquestionably fallible, even if inspired in part by religious devotion. And so this code of law would be designed to change by due process. There are many examples, including the twenty-seven amendments and the many consequences of the Doctrine of Implied Powers. The Federalists had already conceded the need for a guarantee of rights, and by 1791, the new government had ratified the first ten amendments—the Bill of Rights. For better or worse, reform has since removed many of the barriers that excluded the common people from directly selecting their legislators and executive.

At the same time, suffrage has been expanded dramatically.

An interesting question remains as to why the Founding Fathers, who were predominantly committed and activist Christians, were so careful as to ensure that the "Congress shall make no law respecting an establishment of religion, or the free exercise thereof. ... " While a sizeable portion of the colonists and Founding Fathers were outspoken deists, agnostics, and atheists (Benjamin Franklin, Thomas Jefferson, Thomas Paine, Ethan Allen, Elihu Palmer, Aaron Burr, and Joel Barlow, to name a few[25]), they were still only a sizeable minority—why would the majority specifically not follow the customs of the centuries and establish the nation on a religious footing? One theory is that the Founding Fathers, Christian or not, were students of the Enlightenment, and as such they had a profound respect for the recent catastrophes in Europe and their causes. Their contemporary Baron Holbach, an Enlightenment scholar, wrote in *Common Sense, or Natural Ideas Opposed to Supernatural,* in 1772:

The idea of this being [God], of whom we have no idea ... would be an indifferent thing, did it not cause innumerable ravages in the world. ...

Theological disputes, equally unintelligible to each of the enraged parties, have shaken empires, caused revolutions, been fatal to sovereigns and desolated all Europe. These contemptible quarrels have not been extinguished even in rivers of blood. ... The sectaries of religion, which preaches, in appearance, nothing but charity, concord, and peace, have proved themselves more ferocious than cannibals or savages, whenever their divines excited them to destroy their brethren. There is no crime which men have not committed under the idea of pleasing the Divinity or appeasing his wrath.

In light of these events of European history, many American founders sought to institute tolerance not simply of various Christian sects, but of non-Christians and nonbelievers as well.

However, there was another phenomenon that contributed as much and likely more to the separation of church and state: each dissenting evangelical sect of the mainstream Episcopal Church feared that if the government decided to support a particular creed, that the creed so favored would not be their own. Better to vote to keep church and state separate than to see a rival become the sole beneficiary of public largesse. This origin of our freedom of and freedom from religion seems rather less enlightened than many Americans like to think it was and suggests a vulnerability if ever those rival creeds begin to think that they should lay aside their jealousies and focus instead on placing a nondenominational deity at the fountainhead of U.S. law rather than a more inclusive and evidence-based secular ethic. If theological sects need centuries to lay aside their differences and rally under what they have in common—a unifying God—then it seems that it may be centuries more before believers and nonbelievers lay aside differences and rally under what they already have in common—a desire to improve society through an inspiring, healthy, and disciplined code of morality.

In America at that time, religion or abstinence from religion would be an issue central to the citizens and to their lives public and private; it would be a rich part of the history of its people. It would not, however, be allowed to be a structural component of the government, regardless of a clear Protestant Christian majority among the Founding Fathers.[26]

The authors of the Constitution created a document that stands today as the fundamental law of the longest lived democracy in the world; it is the defining entity of the democracy of the United States of America. The U.S. Constitution has been a model of balanced government and an inspiration for the constitutions of many other nations; many of those nations are confident that they have evolved the instrument to a yet higher level.

A.D. 1859: Charles Darwin published *The Origin of Species*. Again, the Protestant Church and the faithful reacted defensively to yet another revelation of the mechanisms of nature previously ascribed directly to God. The Catholic Church was more circumspect. Despite the vehemence of the arguments and the breakthrough in science, there was no breakthrough in philosophy—since Newton, rationalism had proven itself. Mysticism now had its effect only where science could not yet directly go or where individuals preferred their cherished notions over empiric evidence of truth: the existence of a soul, an overseeing divine personality, a final judgment, an afterlife, and a divine plan to the uncontrollable events in life.

While manipulative superstition persisted into the twenty-first century, it is from the late 1700s onward that, in most of Western Europe and its colonies, active resistance to the Enlightenment was waged in a new way: through a culture suffused with proud anti-intellectualism.

In a sense, the "information revolution" had already begun, and as the volume of new information grew, so did the proportion of the population that was left behind by it, bewildered, embarrassed, exhausted, and defensive. To learn from new and complex ideas, one must have the time and resources to be literate, to study the ideas, to discuss your reactions to them, and to have the courage to think and do differently from one's own family and immediate culture. These are resources that few people had. In a predictable retaliation, they declared that all this "schoolin'" wasn't good but bad and they insisted that such "learnin'" was associated with whatever else seemed bad to them: physical weakness, degeneracy, and immorality.

It is a law of the jungle that the learned generally have more opportunity to make more money and to live more comfortably apart from the less-educated masses. But now the size of that more educated and affluent population had grown large enough to be constantly visible, and the obvious cultural differences were also attacked. In the past, delayed and limited childbearing, appreciation of fine arts, and participation in expensive sports were considered contemptible and selfish if not outright immoral. Above all, speech and ideas that suggested broader perspective and that seemed to draw from diverse sources of academic knowledge were the clearest signs of arrogance and impracticality. A man wasn't a "working man," or even a "man," unless he worked with his hands and spoke only of the common fatuities.

America as a nation was being forged exactly during the time of the Enlightenment in Europe. The new nation was founded by a startling group of philosophically minded intellectuals so learned in ancient history up through contemporary politics that they wanted to establish a state with all the best ideas of the Enlightenment. Some even wanted to make the national language Greek so that all the citizens could read the ancient texts in the original form. Of course, they accomplished the former, but the young nation barely survived infancy, and the Greek language never took hold for the obvious reason that the prevalent language of society and commerce was English.

In the general population, however, America bore a heavy burden of anti-intellectualism, for many reasons. As a colony, it was a refuge of the desperate masses; it was a dumping ground for criminals and debtors. Long before the words were emblazoned on a plaque, America proudly invited the "poor, huddled masses ... the wretched refuse" of Europe. There was a self-selection process whereby the Europeans who moved to the American colonies and early states were culturally different from those who stayed in Europe: the wilderness of the new continent appealed particularly to Europeans who felt less attachment to the history and stimulation of the culture of Europe and who had little to lose with such a daring move. The early Puritans, unique in their combination of scholarship and religion, could not keep hold of the masses that had no other intellectual exposure and that continually fled deeper west and south, away from books and debate. The congregations embraced the emotion and accessibility of evangelists, who ridiculed education in any topic outside of their own homespun fundamentalism.

The colonies supplemented this cultural stock with the boatloads of persons kidnapped in Africa, stripped of almost all cultural knowledge, tradition, family ties, and self-sufficiency, and enslaved. The institutions of religion, business, and education developed explicitly anti-intellectual orientations. After John Quincy Adams, the intellectual credentials of the presidents and elected officials would decline in accordance with the demands of the voters.

Intellectualism and reason would discover they did not necessarily have a ready, natural place in democracy, generally or in American culture specifically.

The history of reason in the West continues, of course, to the present day. However, the modern developments are beyond the scope of this book. But if "ontogeny recapitulates phylogeny,"[27] that is, if the development of the individual must retrace the general development of the species, then most of us will be challenged to meet the standards of 2,200 years ago, much less of 200 years ago. The most general and inclusive lessons from those centuries on how to reason well is what we shall examine in the next chapter.

Chapter Two

The Landscape of Reason

There are many causes for our failure to reason well, throughout the history that we have reviewed and in everyday life today. If we are to avoid George Santayana's ominous warning that "Those who cannot remember the past are condemned to repeat it," we would do well to see what, specifically, people tend to do that leads to persuasion not by merit of evidence but by babble, bias, and hype. What follows are the most broad and most pervasive causes of failure to reason well. Subsequent chapters will address more specific pitfalls.

Disdain for Disagreement

To dissent is no offense; it is inevitable and a fundamental right, the exercise of which is vital to a strong democracy. One hopes that the dissent is thoughtful and well informed. If so, it is a healthy challenge that should be made with due process and met with a thoughtful, well-informed response. The person who criticizes a thought or institution may be perceived as attacking it, hating it, wishing for its demise, but may well have deep attachment to it and wish to reform it for its greater vitality. Thus, disagreement need not lurch into contempt or social fracture. Dissent and disagreement are natural and healthy; how we'll approach them with an objective assessment of valid evidence is a critical measure of our civilization.

Confusing Validity with Concurrence

Does a particular idea really lack merit, or do I just not like it because it doesn't agree with my current opinion? Having decided an issue in one's mind may create a defensive fortress that allows in as a "friend" only those suggestions that concur and which excludes any contrary evidence the moment the discrepancy is detected. Such an approach is measuring all ideas with the wrong yardstick— one that measures *agreement* with foregone conclusion, not the inherent *validity* of ideas. This mistake is easy to make for those who do not have a method for evaluating the inherent validity of an idea or those who are comfortable in the certainty that all their current opinions are the only ones that have merit.

Anti-Intellectualism

Anti-intellectualism is an attitude of contempt for the methods of intellectual inquiry and often for the people that ascribe to them. Anti-intellectualism is a powerful cultural force as it is pervasive in many critical institutions of life—family, education, business, politics, and religion. At an early age, some students may discover that they are teased or alienated for having curiosity and enthusiasm for certain subject matters in schools. This anti-intellectual headwind may persist throughout one's education, even at the most reputable institutions, making academic achievement even harder than it naturally is. Some schools and teachers that try to serve as role models for the intellectual life may be so undersupported that the students get the impression early on that education is not particularly valued in greater society. Families may trust too much that the schools can adequately educate their children without substantive conversation, reading, creativity, and other intellectually stimulating activities at home, so any spare time and energy at home slides into vacuous television, video games, and other mentally anemic activities. Politicians may have to swallow their thoughtful or philosophical contemplations so as not to irritate an audience accustomed primarily to lurid topics and bombastic exchange. Many politicians firmly believe that intellectualism is a liability in public life. Religious institutions may cultivate a mindset that accepts superstition and spiritual "feelings" as a valid source for objective answers, and they may be psychologically coercive against, if not openly hostile to, freedom of thought, unrestricted inquiry, scholarship, and dissent.

Business culture and intellectuals have an uneasy interdependence in which the former is dedicated, perhaps excessively, to practical and immediate payoff, which affords little time or patience for broader cultural considerations of their work, and yet they depend on intellectuals for much of the critical innovations that drive industry. Intellectuals may sneer at the makers of "widgets," yet are often dependent on the business community to develop ideas into marketable products and to provide the economic engine (and often the hard cash, through philanthropy) that is essential to academia and to their lives in a vigorous economy. The business world and the academic world keep to themselves except when mutual interest draws them together; even then, the culture shock of one entering the other may be as surprising and unsettling as when any Peace Corps volunteer enters another culture. The gap is attributable to an anti-intellectual attitude of each culture toward the other.[1]

Is the Goal of the Discussion to Discover the Truth or Simply to "Win" the Debate?

These are two very different things. When some people disagree on an issue, they are sincerely trying to work out the validity of the issues involved and are open to any final conclusion, so long as it is the most reasonable one. They discuss everyone's ideas with the intent to find the basic premises that can be agreed upon, the precise points of divergence, and the specific reasons that cause that divergence. These persons ascribe to the philosophy "Only a fool and a dead man never changes his mind" (anonymous) and are willing to consider all evidence and reasoned argument. Engage this person in thoughtful conversation at every opportunity.

Other persons have already decided what they believe and engage in discussion only with the intent to overwhelm the other; impress friends with their clever ideas, little-known facts, or rhetorical dominance; force a favorable vote; hear themselves speak impressively before an audience; score a convert; or verbally defeat or exhaust the other. Whether they admit it to themselves or not, such persons are more of the mindset of "I am willing to discuss it all day until you see it my way." If you expect a thoughtful discussion of your disagreement, you will be disappointed; *avoid* conversation with this person at every opportunity. Unless your interest is the sport of verbal boxing, you will only waste your time and energy with this egotist; hence the folk wisdom "Don't wrestle with the pigs, you'll just get muddy and annoyed, and the pigs will enjoy it."[2]

But most of us grant ourselves special status—we consider ourselves open minded to new ideas and argument; we simply haven't heard, in a long time, any ideas or argument that were worthy of changing our minds. It seems to us that it is others, and not ourselves, who are closed minded.

Public discourse—including many of the merchants of outrage in talk radio, television, and policy periodicals—render the notion of civil conduct itself quaint, if not foreign. Nonetheless, diplomatic conduct when feeling challenged, exasperated, or personally attacked remains a necessary skill of fruitful debate. Patient listening, genuine consideration of the other opinion, and perceptive detection of the origins of the divergence can elevate the discussion without elevating the volume.

Regressing to snide or sarcastic retort, hostility, raised voices, defensiveness, stonewall withdrawal, or sneering passive-aggressiveness are examples of an abandonment of mature effort to trace the disagreement upstream to discover the original points of divergence.

An open mind will find most ideas that are diametrically opposed to one's own ideas not ridiculous or offensive but fascinating, if only because they are so passionately held, and also because perhaps this is the time that the person who holds them will give the evidence and reasoning that you have thus far not seen.

> *I have made a ceaseless effort not to ridicule, not to bewail, not to scorn human actions, but to understand them.*
>
> —*Baruch Spinoza, 1632–1677*

"I Don't Know."

This is often a valid, useful, and very respectable answer. Apart from simply being the honest answer in many cases, it is often the mark of a well-informed and insightful person, not an ignorant or lazy one: scales heavily laden on both sides may not convincingly tip. A scale that is virtually empty but for a few more readily available facts (or worse, token justifications) on one side than the other may clearly tip, but such an answer is unreliable, as what matters is the *breadth and depth of the data* underlying both sides of the issue.

The failure to say, "I don't know" creates many problems. To accept and assert unscrutinized answers creates error, bad decisions, persistent misunderstanding, and personal and historic catastrophe. Doing so is akin to raising an inaccurate road sign: the answer misleads; it causes us to lose time when we should be figuring out the right way to go. It becomes a familiar and established notion that may be very difficult to dislodge. Confucius, circa 520 B.C.E., knew that to say, "I don't know" is a far better answer than any impressive argument that is, in fact, inadequately informed or logically unsound: "When you know a thing, to hold that you know it; and when you do not, to admit the fact—this is knowledge."[3]

There are many reasons we are tempted to create a false answer rather than simply saying, "I don't know" or "I don't know enough yet." A decisive answer is more immediately satisfying than an equivocal one; it gives relief from uncertainty. One's culture may respect decisiveness even more than validity, so we enjoy the social rewards of having explicit answers. The answer we make up, not surprisingly, is often the answer we wish were true, so there is the satisfaction of being right. Answers erected today shall be standing tomorrow; they persist by default, allowing us to move on to other things—all this, so long as we can avert our eyes from the significant possibility that we are wrong.

Parents, for example, may worry that a child will be distressed without easy and pleasing answers so they may succumb to the temptation to make up elaborate fictions to assuage them. Both parents and children are unaware of how deeply the errors will be imbedded, incorporated into one's identity and pride. Grown to adulthood, they may sincerely believe that which, if they themselves were to come from any other childhood tradition, they would glance at and immediately consider patently absurd. So we may enter adolescence certain that one's nation, one's tradition, one's culture is the best in the world, that physical laws have been suspended at the wish of fantastic deities.

Then we may go to college and, for a time, our insight about how much we really know may grow even worse. The sudden rush of new knowledge is so thrilling that we are easily seduced and we are soon accepting and delivering attractive ideas without giving them appropriate scrutiny. We may find ourselves subject to charges of "being sophomoric." Why is "being sophomoric" an insult that suggests you may be publicly humiliating yourself? Because it is an accusation that you've mistaken your superficial knowledge for a complete understanding entitling you to pontificate on complex matters. Second-year college student or not, a little knowledge, a few facts, some vague impressions are all some people need to cobble together a rousing and forceful speech that delivers an opinion with great conviction. Do not mistake passion and certainty, stirred with a few impressive facts, for erudition. The thrill of new knowledge that the sophomore may feel is a good, valid, and important pleasure, but it is the sign of enlightenment *growing*, not enlightenment *achieved*. One hopes it will inspire the student to seek more of this pleasure, to persist in studies, not to leap to excessive confidence or to launch immediately into decisive action.

After school, in the adult world, there remains a preference for ready and confident answers. As a result, there is no shortage of answers in the world, but a marked scarcity of correct ones. To offer an answer where others are stumped will bring you a great deal of attention—"Here's a surefire way to lose weight without all that diet and exercise hassle"; "This is how the world came to be"; This is why there are so many languages"—and at so little a cost of research and deliberation.[4] This is how we get ideas such as the stars are shining human personalities imbedded on a black sphere that rotates around a flat Earth; a misbehaving child must have Attention Deficit Disorder; the secret to stock market success is to time your jumps into and out of the market by my special formula. ... Thus we build a raft of passionately held certainties without examining objectively the evidence supporting their construction.

> *The great obstacle to discovering the shape of the earth, the continents, and the ocean was not ignorance but the illusion of knowledge. ... People who could agree on few other facts about remote regions of the earth somehow agreed on the geography of the afterworld. More appealing than knowledge itself is the feeling of knowing.*
>
> —*Daniel Boorstin,* The Discoverers

The satisfaction of having an answer is not to be underestimated. State leaders take very seriously the social stability that comes with a public that has answers generally agreed upon, regardless of the injustice or inaccuracy of those answers. (See "The Appeal of Social Order" on page 179.)

So in the face of all these temptations to leap to a confidence that is not supported by the evidence, what can we do? We can admit we do not know answers with confidence. We remain curious for the right answer, whatever it is. We develop the skills for finding the sources that tend to produce information more likely to be valid. Whatever information we get, we keep an eye out for admissions of uncertainty. Such admissions often increase, rather than decrease, the credibility of the source, depending on circumstances. Does the history book ever say, "Historians really have little evidence of what happened at this point ... "? The absence of regular "I don't knows" is a red flag for any source. Hold yourself to the same standard. Don't pretend to a substantiated argument if you don't have one.

There is a real advantage to answering questions with "I don't know": it is a natural challenge to find out. Don't shrink from the challenge; go find the right answer. So, in any private deliberation or public debate, identify the points of uncertainty and then do the legitimate research to fill in the gap:

- "I don't know. Show me how many men you think Sparta has."
- "I don't know. Show me why Rome should care if we execute this Jesus guy."
- "I don't know. Show me why there can't be a single-wall hull on the *Titanic*."
- "I don't know. Show me what our national interest in Vietnam is anyway."
- "I don't know. Show me data on how gun laws affect suicide and murder rates."

In the contemporary world, knowing how to do the research is an essential skill that must be learned. One must study and practice the use of myriad databases, physical sources, human specialists, and other resources and know how to filter an ocean of information down to the exceptionally well-substantiated and balanced evidence. Only then do we begin to earn any degree of confidence in an answer that may or may not be definitive.

Personal self-satisfaction is the death of the scientist. Collective self-satisfaction is the death of research. It is restlessness, anxiety, dissatisfaction, agony of the mind that nourish science.

—Jacques-Lucien Monod[5]

The fallacies of reasoning in the following dialogue may be obvious, but can you identify them by name?

Dave: I can't believe that that woman golfer would challenge the PGA Tour on its exclusion of women. It's ridiculous and contemptible that she would even go public and call it unconstitutional.

Maureen: Why is it ridiculous and contemptible?

Dave: Because it's the law. Because a private club can do what it wants to. If the club were taking public money, that would be different—but this is a private club.

Maureen: So a person can open a private club for pedophiliacs?

Dave: No, of course not. That's a completely different thing.

Maureen: What makes it different?

Dave: Little kids are getting hurt. No one gets hurt by expecting women to have their own golf club. I'm just glad that the courts had enough common sense to uphold the law.

Maureen: I'm not sure "hurt" has to be physical—it can be economic too. It can be freedom to participate equally. What was the court's reasoning? What was the law that you say they upheld?

Dave: I don't care. I just heard that they decided against her, and I'm glad they got it right this time.

Maureen: I'd like to know the reasons they gave. I think it's hard to

There is an important historic analogy between the human difficulty in saying, "I don't know" and the human difficulty with the concept of zero. As self-evident as it seems today that zero would have its own place on the number line, the historic reality is that for centuries, humanity, including scholars and mathematicians, did not recognize that "no amount" was as valid an answer as "this tangible amount." To them, it seemed that if you were speaking of an amount, you obviously had to have an amount to show. In effect, they were trying to do math with only the digits one through nine. Although basic arithmetic and even complex calculations were being attempted by Sumerians more than 5,000 years ago, they—and all known cultures for thousands of years, including the Greeks, the Hellenistic cultures, and the Romans—were continually confounded and led astray in their calculations and could not figure out why. Not until A.D. 873 did any record show that the notion of zero and a symbol for it was finally utilized as a placeholder and as a valid amount alongside other quantities in calculations. (This profound contribution came to Western culture through the Arabs, but probably originated in

India.) By allowing zero to be as valid as one, two, three, and so on, there was no longer a tendency for other numbers to slip into place where a zero rightfully belonged. Thus, the maddening tendency for an otherwise valid calculation to produce an absurd answer or an answer misleadingly believable but in reality wrong dramatically evaporated.

Analogously, we desperately need to employ "I don't know" in our nonmathematical reasoning on complex social issues. "I don't know" is a valid answer and an appropriate placeholder. As long as we have an aversion to "I don't know," we will allow incorrect answers to slip in, and thus our reasoning will remain as confounded, and our final outcomes will be as erroneous, as the mathematics of the ancients.[6]

distinguish between having exclusive clubs to keep out racial groups versus keeping out women or keeping out anyone else—what constitutes a "protected class" anyway?

Dave: Maureen, you've had your nose in the books so long that you have lost your common sense. You need one less college degree and a little more street sense. You are exactly the type of Orthodox Liberal that gives Liberals a bad name, reflexively parroting any old lefty dogma.

Equal Evidence, Equal Merit

How much valid material evidence do you have to support your opinion? An opposite opinion with the same amount of valid material evidence deserves the same amount of respect. Your idea merits adoption no more than the other. This is certainly *not* to say that all differing opinions are equally valid; the whole point is to develop a precise and fair method of weighing the relative merit of each idea and to make valid judgments regarding which carries the greater value.[7]

A very important corollary to this is that if you claim *no* valid material evidence at all for your opinion but assert its truth as "just a matter of faith," then, since what is allowable to one side is allowable to the other, the opposite claim is likely to be asserted on "faith" alone as well. You both claim zero material evidence, the scales do not tip at all, and, so far, you have only equal merit (or equal lack of it). If there is to be a reasonable choice made, then each side will want to find valid *material* evidence to justify his or her position. If at least one side offers concrete real-world evidence, which can be scrutinized against known science, history, economics, and human nature, and so on, then we can make a valid choice. But what was originally asserted on faith alone must always be backed up by concrete material evidence, or it is simply negated by an opposite claim based on faith alone.

Human conflict frequently provides examples of persons who admit that they cannot demonstrate the superiority of their assertion but who yet insist, based on their faith alone, that it be adopted, even legislated, on those who disagree for reasons equally or more valid. Such legislators are, they may have us believe, especially exempt from having to demonstrate the material superiority of their position. (More on faith later.)

Does the Question Even Have a Right Answer?

There are several ways to get lost in your search for the right answer. Three that are closely related are discussed in the following paragraphs.

The first error is in confusing *objective* questions with *subjective* questions. What type of music is good music? Do rutabagas taste good? These are *subjective* questions because the answer depends on indefinable notions of taste, preference, and values. Sometimes, but not always, this means that any person's answer is as valid as the next. In contrast, consider these: does the sun go around Earth or does the Earth rotate? Does Santa Claus or Vishnu actually exist? These are *objective* questions for which reality has a specific and correct answer, whether anyone on the planet knows it or not, regardless of sincere error, daunting tradition, or cherished preferences.

The second error is in confusing a disagreement in fact with a disagreement in fiction. Fiction is even less rooted in the real world than subjective issues. What does Mrs. Claus feed the reindeer? What does the Greek god Dionysus think about cloning people? One answer is as good or bad as the next because we are making it all up anyway. One answer may correspond to the established myth better than another may (Santa's suit is *red*, by golly), but there is still no correspondence to reality.

The third error is in confusing fact with truth in its larger sense. The events in the fable about the tortoise and the hare may not have ever actually occurred per se, but the "truth" represented therein about persistence and patient progress is no less valid.

Each of these is explained a little more thoroughly below.

Claiming That an Objective Question Can Be Answered Subjectively

An *objective* issue has a specific, concrete answer in the real world, whether the answer is known or not: Has the e-mail I sent been read yet? Is the moon round as a ball or flat as a wafer? Is there a living *Tyrannosaurus rex* anywhere in the world?

A *subjective* issue does not have a specific reality-based answer, but is subject to preferences and uncontrollable conditions: What is the best color for a bedspread? What is a better motivator: hope for reward or fear of punishment? How can I act now to best show kindness, sensitivity, and moral decency? Subjective questions are difficult to answer through scientific analysis because they are subject to preferences and uncontrollable conditions. There is more than one "right" answer or no ultimately "right" answer at all, because the question is subject to many vaguely defined variables. Thus, persons are free to answer in any way that seems reasonable to them. *This freedom is so pleasant and convenient that it is tempting to claim it when trying to answer objective questions,* but that would be an attempt to dodge the responsibility of objective scrutiny for an objective claim, or it may be an attempt to avoid or deny an unpleasant or less preferred objective reality. Abe Lincoln is credited with making the point thus:

Question: How many legs does a dog have, if we call a tail a leg?
Answer: Four. Calling a tail a leg doesn't make it a leg.[8]

Would you call each of the following questions objective or subjective?

1. Does smoking cause an increase in rates of disease and early death?
2. Did charter schools improve overall student performance on standardized tests in Denver County last year?
3. Does compulsory military service make young people more responsible?
4. Does abortion always offend God?

1. This is an objective question because it has a specific answer in reality (in addition, it is very helpful that we have discovered that answer and can choose to act accordingly). However, we have all heard truth sacrificed for preference or "to keep the peace," as if this objective question were subjective: Does smoking cause disease? "Well, if a person really wants to smoke and does so of her own consent and can find at least one study, no matter how scientifically flawed, to say that smoking is not associated with disease, then we must grant that her answer is as good as any other." No, we do not have to grant that, because it is not a subjective question open to more than one correct answer—objective claims are inherently at the mercy of objective scrutiny. If the data were truly ambiguous, then we would need better data, but it is no less an objective question. Only truly subjective claims enjoy the freedom of subjective interpretation.

2. Ditto. This is a specific question with a specific answer in the real world, whether we have done the necessary research to find the answer or not. If the question had been a more vague "Do charter schools make students better citizens?", then the question would be subjective because of the

ambiguity in what a "better citizen" might be.

3. This is subjective, because what makes a person "more responsible" is defined quite differently by different people. A strict definition of responsibility and how it is measured would be necessary for this to approach objectivity. Disagreement as to the answer, even for a subjective question, does not necessarily mean that both answers are equally valid—some are better supported than others by reason in the forms of science, history, psychology, and so on. That is, some subjective answers are still inherently more "reasonable": Do automobile safety regulations save lives? (objective question); Are the lives saved by those regulations enough to justify the infringement of business freedom and increase in cost of the cars? (subjective, yet with a compelling answer).

4. This is objective. Whether God exists or not is an objective question, regardless of our preferences and certainties. Similar to the Loch Ness Monster and the book you hold in your hand, He/She/It exists or does not; whether we can prove it and know the answer is another issue. Even if we knew He existed, the claim that God is always offended is also objective—He is or He is not always offended, but we still have to find out, if we can, what, in fact, His opinion is.

Any objective claim, such as a claim that an entity exists, is a claim of knowledge of real-world actuality and is therefore subject to objective scrutiny. If it is said to exist, there must be a preponderance of legitimate evidence that it exists, which cannot be reasonably explained otherwise—for example, that it is a fabrication to serve some practical or psychological purpose. Alien abductions, angelic visitations, previous-life possessions, the economic effect of a legislative bill, and a news story are all claims of objective existence or event. And this brings us to the next distinction: objective reality versus outright fiction.

Clarifying Reality and Clarifying Fictions

Does the ghost-spirit of a Native American tribe prefer corn or meat as a sacrificial offering to induce rain? What does Harry Potter listen to on the radio? If you accept the premise that these entities exist, then they are objective questions. But let us back up—many would say with confidence that these entities do not exist, so these are questions that can have no correct answer in the real world, since they are based entirely on fictional stories. If they have answers, then those answers—similar to the stories on which they are based—are entirely subject to the discretion of the human authors who created them. Persons who debate issues of fiction, especially those fictions that deal with serious human issues, may *feel* as if the subjects are based in reality, they may *want* their answers to have an objective basis in the universe, and they may *insist* that knowing the answers is achievable through "gut feelings"—but none of this changes the fact that what they feel and want and insist upon is fictional.

Another example: many persons would like to think that the Loch Ness Monster is real, but all "evidence" for it is explained more credibly without having to resort to presumption that there is a fantastic creature living among us. Furthermore, the belief that there is a Loch Ness Monster is likely perpetuated by interested parties who make money (off the consequent tourism), or who enjoy status (by being an authority on something fantastic), or who offer comfort (by fulfilling a psychological need for something mysterious, exciting, and powerful for us to be in awe of). However, when the supporters of such claims are asked to show legitimate evidence or describe how such is possible, the retort is sometimes that such a request is meaningless and naïve because such things are beyond science or evidence or reason. No, they are not, precisely because they are objective claims. *The thing exists or it does not.* There may be inadequate data at any given time, but this means only that their claim remains a speculation or a wish or a comfortable delusion. But such a notion is not evidence; it is the call for evidence. A desire for something to be true is not evidence that it is true. There is no valid evidence that it exists, but there is ample valid evidence in human psychology and economics as to why some would like to insist it exists. If the preponderance of objective evidence indicates that the claim is false, then reason dictates the dismissal of the claim or at least a call for more valid evidence.

Perhaps in the future we will have some direct revelation or we will invent instruments that will give us the convincing evidence that has eluded us thus far. Until then, the existence of the Loch Ness Monster remains a delightful and useful fabrication that pales under the weight of all the evidence, physical and psychological, that it does not exist, except in the hopeful minds of the faithful who prefer a world in which there is a Loch Ness Monster to a world in which there is not.

Confusing Truth with Fact

Language is a clumsy tool. Psychologists and linguists have long recognized that language can powerfully dictate what concepts we "see" and "don't see." The traditional Inuit of the North are said to recognize dozens of different types of snow (perhaps because the types of snow so greatly affect their transport, livelihood, and so on), whereas many of us, while experiencing the same kind of weather, would see only snow, hail, or sleet.[9]

So, too, with "truth." In modern language, we are led to believe that an event is "true" only if it actually happened as described literally. But the story is often a fable, as in the examples above of the Inuit or of the tortoise and the hare, or is a similar vehicle to teach a particular *truth*. Because we use the same word, often we do not see the difference in the concepts.

The Education of Little Tree, a fictionalized version of the youth of Forrest Carter, may challenge some people's preferred notions of the world, but it reflects a great deal of truth about human nature. Too many people, rocked by the concepts therein, retreat to a sole criteria: "Is this true? Did this really happen?" And if you begin to answer, "No, but the *idea* that a person can be happy even if he chooses to have fewer material possessions … ", your friend may be already tossing aside the book with, "Whew! I knew it wasn't true!" Your friend loses out.

Many books claim to be literally, objectively true. They may be or may not be, but there are lots of reasons why that claim might be made nonetheless. Sometimes literal truth matters a lot, such as instructions on how to operate your parachute, conduct a science experiment, or in any book in which you are hoping to find explicit statements of fact. But sometimes the particulars matter far less than the underlying concepts or truths therein. If you are looking for general truths about human nature, for example, a great deal of fiction, parables, and even ancient books that may not be historically accurate in every detail can be very truthful. Are Plato's *Dialogues* historically accurate? Did the conversations with Socrates actually occur as stated? Ask yourself if historical precision really matters to you as much as the overall message and implications of the work. Simplistic thinkers will get hung up on whether something is literally true and will pass judgment on its value and relevance by that test only. Sophisticated thinkers can learn to discriminate when literal truth matters and when it does not. The confusion starts early:

"Daddy, is that true? Is Santa real?"

Deer in the headlights, Dad is thinking fast. … Well, when we think of this kind man in a red suit and something happens in which children are inspired to behave better than usual, and then there are presents under the tree, … the phenomenon as a whole seems to happen as they say. … So, in an abstract sense, … "Yes, honey, it's true—Santa is real."

If we do not learn early on to distinguish what is literally true from what is "true" in an abstract sense, after a great deal of interpretive massaging, then we can become too accustomed to finding literal truth in flagrantly absurd claims. A discerning reader may find a great deal of truth in books known to have errors of fact and a great deal of misinformation in books that claim to have a lock on objective fact. Whether a book claims to be fiction or nonfiction, infallible or fallible, is an inadequate criterion for finding books that contain valuable truth.

Home Team Bias

Being ignorant of foreign ideas and ways makes problems appear simple and clear. Knowing much about my own gives a sense of thorough understanding. From a distance, big things look small and very different things look indistinguishable. It is in human nature to consider something small, distant, or unfamiliar as being insignificant, irrelevant, or inferior compared to the things at hand. If it does not affect your life now, then you need not worry about it, right? Perhaps. But you may also miss out on many valuable ideas and experiences—and warnings. Worse, you may adopt certain ideas and occupy yourself with certain activities that you yourself would consider far inferior had you previously gone out of your way to expose yourself to those other ideas and activities. As a matter of ego, each of us would like to think that by birth we happened to land into the best of worlds—no other country, cuisine, constitution, breadth and depth of education, music, religion, or ideas could be much better or more enriching than those we have found handed to us. But since there is a tendency for most people all over the world to believe that, perhaps it is no truer for us than for all the others. It is the Home Team Bias. It is natural to have a warm affection for the football team, the culture, the constitution of the world in which we are raised; few persons upon reaching maturity can set aside those attachments and choose freely and objectively by relevant criteria.

If I decide that the ideas of my family heritage are indeed those that I consider to be the most worthy *after* I have set aside my cultural heritage and put equal effort into identifying, reading, discussing, and understanding the others, then it is a solid and informed decision. Otherwise, one runs the risk that the ideas one holds passionately—the religion, the politics, and so on— are just accidental outcomes of birth, history, geography, and previous conquest. Those persons would believe with the same degree of certainty the ideas of any other time and culture had they been born then and there instead. If you can look around and see that you share the same ideas as your family and immediate culture, you should look closely to see if ideas are absorbing and using you rather than you selecting, absorbing, and using ideas.

Self-Interest

It is a matter of human nature that we are influenced by our own self-interest. We tend to place self over others, our family over other families, our culture over other cultures, and our species over other species. It may be understandable, but in the context of the greatest good for the greatest number, is it rational or is it frank bias?[10]

Self-interest can influence greatly what we *choose* as sources of information, because many people do not seek out opinions that will challenge their current opinion—they simply do not want to be troubled or annoyed. Self-interest can influence greatly how we *interpret* the data that we are exposed to, viewing it in a light that assures the interpretation accords with our present beliefs, preferences, and benefits.

> *Deliberation. The act of examining one's bread to determine which side it is buttered on.*
> —Ambrose Bierce, The Devil's Dictionary

Few people are able to view issues from the perspective of someone else's interest: What would I think about this if I were poor? African American? Differently educated? Not that the other person would be any less influenced by self-interest, but do the other perspectives reveal any considerations unaddressed by one's own particular perspective? Would I feel the same way if the idea caused me to have more money rather than less? To have more power/attention/sex rather than less? Am I deciding this according to fairness, consistency, and the greatest good for the greatest number or according to what works best for me? If it's for me, is that okay?

Entire worldviews and philosophies may be constructed around one's self-interest. Anthropocentrism is the human tendency to place humanity in the center of all that is meaningful, physically and morally. It interprets the existence and purpose of all things as being directed toward mankind. We do not do this because it is justified by any evidence; perhaps we do this because it is easy, self-aggrandizing, and reassuring. Anthropocentrism was evident in our assumption that the sun revolved around Earth and that all the other planets and stars did so as well. Some traditions, though not all, believe quite literally that humanity was modeled by Universal Omniscience in His own image, that humanity was given the world as his domain to name and to use as his own "resources." Often, where men were the exclusive authority, the supreme god was a male. Natural events are rewards or punishment, lessons or tests, from a god who is intimately concerned with humanity as a whole and every individual. According to this worldview, the chief interest and intent of all existence is directed at our species.

One need not be religious to conclude that mankind is the center and summit of the universe. The truth of this seems self-evident to us as we look at our intelligence, our mastery over our environment, or our place on a stage in the center of revolving heavenly spheres. We may forget to factor in humanity's adverse impacts on long-term ecosystem viability or some of the savage wars we have inflicted on one another. We may forget our relative size in space and time. Not everyone can escape the suspicion that we judge by standards of our own choosing; if the nonhuman parts of creation could contemplate the question, they might judge not by use of tools and technology but by other standards: cockroaches might judge creatures by resilience through the eons; germs by numerical superiority and subtlety of power and relentless subjugation of humans, individual and global. All nonhuman animals might judge not on the ability to manipulate the world to serve their own needs, but on the ability to live in it in an ecologically balanced and sustainable fashion. Individuals can choose to believe that which best serves one's preferences; humanity in general can do so as well.

If we try to imagine a worldview unbiased by the self-interest of anthropocentrism and better supported by broader considerations of evidence and history, we may instead reason that the cosmos operates by physical law and never has and never will violate physical law; that from the fusion reactions of stars, all elements stream into the universe and form the structures present; that our species is one of many that have arisen naturally, without intentional design, in a complex web of species and ecosystems; that in pursuit of our self-interest and self-indulgence, we could tear at the fabric of the web at the peril of ourselves and of all species; that we find ourselves adrift on an obscure mote of rock trapped by an ordinary star, which is one of hundreds of billions that trail on an outer edge of a typical galaxy, which is itself one of hundreds of billions; that all humanity occupies an infinitesimal place in time as well as in space; that we exist through constant flux of atoms and molecules into and out of our bodies in an exchange with the chemical milieu around us; that we are not given meaning and purpose as an unearned birthright, but that through individual and collective effort we achieve meaning and purpose, or fail to do so; that, eventually, each of us dies, and our consciousness, being only a product of our physical brains, dies with it; that with death, we cease to exist, but for the atoms of our being that disperse by the same uninterrupted and unaware chemicals laws that had previously built and sustained us; that in about 7 billion years, even Earth's inner solar system will cease to exist as we know it when our star becomes a red giant, destroying all life on Earth, and perhaps the Earth itself, before retracting into its final form as a black dwarf; that all the while, other stars and solar systems also are born, exist, and die.[11]

From ashes to ashes indeed.

Some readers may find this worldview offensive or disheartening, while others may find it beautiful and inspiring. But the aesthetic of the worldview is not the point; the point is that this may be the worldview that best accords with the valid evidence, considered without self-interest.

Thomas Hardy felt no loss when he rejected anthropocentrism in writing:

Let me enjoy the earth no less
Because the all-enacting Might
That fashioned forth its loveliness
Had other aims than my delight.

In critical analysis, we must be aware of our biases, including our tendency to be biased in favor of ourselves, our family, our social group, our nation, our species, and so on.

Self-Pity

"I can't believe that—it seems so bleak." Sometimes we shrink from reality. We have to believe a reassuring notion because we cannot bear to believe otherwise. It is a forgivable reality of our human nature that we often cannot hold up emotionally against all facts of this world. Forgivable, perhaps, but nonetheless, we are choosing to believe that which is not true.

Sometimes we choose the answers that appeal to us the most and spare ourselves the unsettling feelings of doubt and questioning, researching, accepting uncertainty, and changing one's mind. Such efforts are laborious and often truly painful, so we take pity on ourselves and choose not to question.

Believe Now, Understand Later

Belief is the tentative *conclusion* of deliberation and comes from the net balance of sufficient evidence; it is not the starting point from which to begin to seek supportive evidence. First, we have questions, then we seek information and carefully weigh each piece for its validity, and then we develop an understanding that we state as a tentative belief. But some would rather leap straight to the belief and skip the messy middle part. Why?

It is the insightful questions, the balance of all evidence, and the careful deliberation that give belief (if it ever comes at all) its strength. Thoughtful and informed persons seldom believe or disbelieve 100 percent of the information they come across on complex issues. Rather, they often believe evidence for both sides of an argument, have many unanswered questions, and so have a sense of lingering uncertainty, even if they generally lean more toward one direction rather than another. So why might some persons or institutions attempt to strong-arm you into declaring a belief, dismissing the fact that you have unresolved uncertainties, perhaps even telling you that your doubts are shameful or pressuring you to declare for others as well as yourself what it is you believe? Perhaps they seek vindication of their inadequate argument, seek to grow the membership of their movement, or seek power for the sake of power. Perhaps they believe that a declaration of belief from you would somehow make it so. Or perhaps they hope that if you declare belief, you will have a sense of obligation to that belief and to the audience that heard you say it. But you need not accept that pressure and manipulation. It is acceptable and perhaps considerably more logical to refrain from declaring a belief one way or the other until you feel you have adequate evidence and have answered, for yourself, all pertinent questions.

Faith and Intuition

Faith and intuition are special types of "thinking" that tempt us to claim that we can understand objective reality or act without adequate objective evidence (and perhaps despite compelling objective evidence). Unfortunately, interesting disagreements about evidence are often abandoned when one side plays the "faith card": "Never mind. This issue is inherently beyond and is impermeable to crass things such as evidence, reason, science, natural law, and common sense; it's just a matter of faith."

What is your definition of faith? The word is used in very different ways. Consider the following:

1. I'll ride this roller coaster—I have faith in the people who built it.
2. Those of us faithful to the communist ideal are unfazed by the disasters of the Soviet Union, China, Cuba, and East Germany and still call for a glorious worldwide revolution!
3. I rub the exotic lamp and make a wish. Sometimes things work out the way I hoped, sometimes they don't. I always get an answer, just not always

the one I wanted. Nonetheless, I won't question my faith that the lamp always hears and acts in my best interest.

4. After exhaustive consideration, this time I put my money in the mutual fund your brother mentioned. Sometimes you just have to go on faith.

There are two general types of faith: conditional and unconditional. In *conditional* faith, our belief or our action is based on a sense of trust, but if evidence begins to accumulate that the trust is not deserved, then the faith is retracted. Consider the roller coaster at the amusement park: there may seem reasonable basis for trust in the engineers and operators of it—or no known reason to withhold trust—so the riders "take it on faith" that they will be safe and climb in. If the roller coaster crashes, they will instantly and unapologetically drop their faith in the builders and operators. Who would call these injured riders who rejected their faith arrogant, ungrateful, somehow deficient in character, or evil? They would probably be called sensible people capable of learning from legitimate evidence. Conditional faith is similar to initial trust or to the optimism of giving a person or an institution the benefit of the doubt until there is reason or evidence to judge otherwise.

In *unconditional* faith, no amount of evidence contrary to one's faith is sufficient to dislodge it. It is allegiance and obedience to the faith itself that matters, not the existence of evidence one way or the other. Evidence *against* the faith is assiduously avoided, ignored, deflected, or skeptically scrutinized far more rigorously than is the evidence for the faith itself. In fact, the believers may be quite proud about being unperturbed by evidence or reason and feel that determinations based on emotion, preferences, uniformity, and social expectation are not only valid but noble.

Can you say which of the above four examples seems to reflect conditional faith versus unconditional faith?[12]

Many institutions include a degree of this type of unconditional, or "blind," faith. Scientific institutions often have some "scientists" and lay members who seem to believe that white coats, expensive equipment, and conventionally formatted publications must necessarily confer scientific validity, and, without adequate scrutiny, they thus produce a great deal of pseudoscience. Blind faith in political institutions can produce mindless orthodoxy and crude demagoguery. Blind faith in religious institutions has left a horrifying swath in the history of many societies. Each of these great institutions may be highly beneficial to society in many ways, but the subgroup within each that believes unconditionally may do more harm than good to its institution and to society. Different communities of scientists may have quite different levels of unconditional faith present, and the rational contribution of each of those communities may vary accordingly; so, too, with each political party and each religion. It is not the institution *overall* that is so problematic, it is the thinking and behaviors within each institution that occur when reason succumbs to unconditional faith.

Unconditional or "blind" faith is the Teflon of the world of evidence. The faithful may walk away from a torrent of evidence, unmarked. Why?

March 20, A.D. 36: The world will come to an end within a few years!

Sometime in A.D. 100: Well, it could be any time now! Prepare thyself!

A.D. 1466, on sighting Haley's Comet: Surely, this is the end!

Throughout A.D. 1999: The world will come to an end on December 31, 1999!

January 1, A.D. 2000: Okay, the message got scrambled—but the world will end soon!

If unconditional faith can be so misleading, why was it not discarded long ago as a reliable tool? Because even if it is not reliable in finding objective truth, it serves other needs. Unconditional faith fosters stability, uniformity, cohesion—in the short-term anyway. It puts to rest difficult questions and irreducible uncertainty. It may be self-serving without having to answer for being self-serving. It may delay a humbling admission of error for another year or another generation. It is often intellectually accessible to the masses and often very compelling. The English language and centuries of tradition endow the word "faith" with positive and noble connotations—it suggests trust, resolve, and strength of character on both the believer and the believed. Sometimes, unconditional faith feels good.

The opposite may be true of conditional faith: it admits uncertainty, acknowledges the possibility of error, opens the door to potentially endless reassessment, invites diversity of opinion, and embraces a healthy change. Adequate consideration of an issue may require a great deal of time, money, study, self-doubt, and stamina. Uncertainty may never be satisfied. Perfect consistency may be impossible. If faith is self-serving, the believers will have to materially justify it. For many people, conditional belief is often uncomfortable; they may claim that it connotes unreliability, lack of commitment, and weakness of character.

While a statement such as "It is not supposed to be rational, it's just a matter of faith" is commonly used to answer an argument and to strike a noble pose, a critical thinker will wonder if it is always so positive and noble when a belief is held steadfastly against all adverse experience and proof to the contrary. When does faith become not a guiding light through dark times, but a blinder from new information, hard questions, and the responsibility to answer them thoughtfully? When does blind faith become a mechanism of consolation or a source of easy answers, however false, or just plain egotism unwilling to admit error? Blind faith may serve as an escape hatch whereby the person ends his own personal deliberations, or ends the discussion, by declaring special exemption from reason and scrutiny and consoles himself for abandoning reason by affecting an ironic air of lofty ideal. Those persons may indeed be noble, trustworthy people of strong character, but not necessarily because they have blind faith. They have succeeded only in putting themselves on the sidelines during a useful intellectual inquiry. Such a stance is assuredly safer and less strenuous than the inquiry, but it does not advance the discussion. "Faith" may be an adequate answer in some situations, but not for those that concern claims of objective reality.

One famous episode of the debate of unconditional versus conditional faith was seen in the life of Galileo Galilei. In 1616, Galileo went public with his scientific proof that rather than the sun revolving around Earth, Earth rotated around its own axis and also revolved around the sun. This contradicted the Church's view of the structure of the universe, to which all the faithful were expected to adhere. Galileo was summoned to Rome and told to "neither hold nor defend this idea of Earth moving around a stationary sun." He continued to do both nonetheless, and so was shown the instruments of torture. He was finally forced to recant and was spared by being put only under house arrest until he died.

Egregious violations of freedom of thought, freedom of speech, and intellectual integrity are not limited to only the remote history of the Church. As recently as the early twentieth century, both Catholic and Protestant churches continued doctrinal opposition to science and continued to compel their members to obey these directives. Attempts by scholars and clerics to reconcile Christian doctrine with recent scientific discoveries had developed a strong following in both the Catholic and Protestant churches. These "Modernists," as they were collectively referred to by Pope Pius X in 1907, tended to deny the literal interpretation of the Bible and to deny the objective nature of traditional beliefs, regarding some rituals as symbolically valuable but not necessarily objectively true—for example, the transubstantiation of bread and wine into actual flesh and blood.

Censure of the movement reached a climax in 1907. On July 3, a decree, *Lamentabili Sane (With Truly Lamentable Results)*, was issued by the holy office with the approval of Pius X. It listed and condemned sixty-five propositions as heretical, false, rash, bold, and offensive; thirty-eight of them related to biblical criticism and the remainder to Modernism. "Lest they captivate the faithful's minds and corrupt the purity of their faith, His Holiness, Pius X, by Divine Providence, Pope, has decided that the chief errors should be noted and condemned by the Office of this Holy Roman and Universal Inquisition." The leaders of this movement knew very clearly they were in the crosshairs of God's chosen representative.

Two months later, the Pope issued an encyclical, *Pascendi Dominici Gregis (Of the Primary Obligations)*. Modernism, it said, is a synthesis of all heresies, "an alliance between faith and false philosophy," arising from curiosity and "pride which rouses in them the spirit of disobedience ... they are found to be utterly wanting in respect for authority, even for the supreme authority."

Pius concluded his attack on modern historical and scientific scrutiny of doctrine on Sept 1, 1910, in *Sacrorum Antistitum (The Oath against Modernism)*. He validated all articles of Roman Catholic belief and rejected anew all the tenets previously rejected by the Church of Rome. He commanded his flock to comply in writing. The document opens, "To be sworn to by all clergy, pastors, confessors, preachers, religious superiors, and professors in philosophical-theological seminaries. I firmly embrace and accept each and every definition that has been set forth and declared by the unerring teaching authority of the Church, especially those principal truths that are directly opposed to the errors of this day. ... "[13]

Yet change came anyway. In 1943, the encyclical *Divino Afflante Spiritu* by Pope Pius XII officially sanctioned modern principles of exegesis for interpreting the Bible. In 1983, Pope John Paul II tried to put further distance between Church history and his own time when he passed over Pius X entirely and said of Galileo and of enlightenment science generally:

> We cast our minds back to an age when there had developed between science and faith grave incomprehension, the results of misunderstandings or errors which only humble and patient re-examination succeeded in gradually dispelling. The Church herself learns by experience and reflection, and she now understands better the meaning that must be given to freedom of research.

This certainly was a dramatic change for a church that had, only seventy-three years earlier, called itself "unerring." Still, others may say that the people running the Church for the last 600 years or more erred by confusing its current opinion with absolute fact—an error many of us still commonly make. To do so is in the nature of people, and this happens in all institutions because all institutions are ultimately "only" human. Perhaps the people running the Church erred by using subjective methods (references to tradition, to the stability of having unchangeable answers) to try to understand the objective realities of the structure of the universe. Perhaps they erred by not being willing to say, "I don't know" when asked if Earth was the center of the universe. Perhaps they let their self-interest and pride outweigh scientific evidence. Perhaps they could have salvaged credibility if they had been able to simply say that they were wrong and had been corrected.

Regardless, some of the lessons here are that all of us must be aware of the dangers as well as the comforts of blind "faith"; to know whether the issue at hand is one of an objective or subjective nature, to accept only evidence that fits that nature, and to encourage humble self-revision as new data necessitates.

By September 1998, when Pope John Paul II issued the encyclical *Fides et Ratio* (*On the Relationship between Faith and Reason*), the Catholic Church's opinion on science and modern philosophy had changed. Now reason was an equal and essential partner: "Faith and reason are like two wings on which the human spirit rises to the contemplation of truth. ... " Indeed, it remains unclear which of the two had been thus promoted into equality; the letter would read as a melancholy appeal that humanity not forget about the potential value of an old, once intimate, still faithful faith, despite the alluring new vivaciousness of science, technology, and modern exegesis. With the passing of Pope John Paul II in April 2005 and the subsequent election of "hard-line conservative" Cardinal Joseph Ratzinger as Pope Benedict XVI, the world waits to see what the new pontiff—and those who succeed in turn—will do with this inheritance of recent change. Patience is advised. Religious doctrine and glaciers do indeed change powerfully and with effect, but they are better tracked in the history books than in the newspapers.

If some faiths have relinquished to science their claim to authority on the objective realities of the world, the license was not granted without restriction. Fundamental claims about the existence and nature of God, souls, and the afterlife are objective claims; thus, they are inherently matters of evidence and rational plausibility. Yet these issues are still held by some churches as being exempt from the rules of evidence. The rationale for the inconsistency is debated.[14]

Self-Censorship

Throughout recorded history, censorship of inquiry, speech, and writing was the norm in institutions private, religious, political, and academic and was deeply oppressive to the development of civilization. Even with the development of the most advanced democratic societies, censorship remains commonplace. Through human effort, however, censorship has receded in many places. Many personalities and conflicts in history that involved censorship are famous—or infamous—and yet they represent only a tiny fraction of those oppressed or executed: Anaxagoras, Socrates, and Aristarchus among the ancients; the Inquisition, Giordano Bruno, and the *Index of Forbidden Books* in the Catholic Church; the constant harassment of Jean le Rond D'Alembert, Voltaire, and the other Encyclopedists of eighteenth-century France. Although revolutionary cultural changes have weakened the existence and strength of censorship in many democracies, today there remains lively discussion of the appropriateness of persistent institutionalized censorship—consider, for example, the current U.S. prohibition of publishing photographs of American soldiers killed in war.[15]

However, the focus on explicit institutionalized censorship may be misplaced. What was accomplished through censorship by self-interested authorities in centuries past is achieved far more effectively today when the minds of individuals are trained to believe that questioning the tenets of one's parents, teachers, country, or faith is arrogant or disrespectful. There is no need to ban books when the people themselves habitually avoid books that seem, at first glance, contrary to their preferred ideas, or if they read only their own scripture and not that of other faiths, or if they subscribe preferentially to those political periodicals that tend to argue the side of an issue that one already believes. Institutions no longer need a thumbscrew or an iron maiden when a sad look of crushed disappointment from a mother or quiet alienation from former friends is enough to convince a person to drop the discussion or to set aside the unorthodox book. If we confuse our *love and respect* for family, friends, country, and faith with our *agreement* with them, then we do to ourselves what the pyre could not.

People's minds are often carefully and explicitly prepared for defense against new ideas in order to preserve a very specific philosophy that is considered immutable and beyond question. Young people get a specific message, even a commandment, to steel themselves against any challenge to the opinion they are expected to hold. The lesson, reinforced and reiterated in a variety of forms over time, usually amounts to something similar to this:

The world is full of persons with ideas different from ours. Whether they are in innocent error or have malicious intent, they will try to pull you away from our true teachings. Be on guard. Good and noble persons, of course, will avoid the people and experiences that carry those corrupting ideas. Yet, in a culture saturated with ignorance, error, and deceit, to have some such encounters is ultimately inevitable. Thus, you must be at the ready, sharp with alertness, as the vigilant guard who stands alone facing into the darkness, watching and listening for the first sign of intrusion. The moment you detect something contrary to our teachings, identify it immediately and shut all possible avenues of entry. Do not be deceived—recognize quickly the words and actions that belie a deceiving outsider—when the hand of friendship is actually a poisonous tentacle, when the wise teacher is, in fact, a devious liar. The ideas may be delivered by a seemingly ordinary, decent, and fair-minded person; the ideas may be crafted to sound measured and reasonable. When you feel such a scorpion on your neck, strike it away and instantly crush it.

If you allow even the slightest doubt in our teachings to exist in your mind, then the deceivers will know it, go to it quickly, and there establish their corruption in you. From that moment, you may be lost—it may be only a matter of time. First, you will know the pain of doubt and uncertainty. Our clear teachings of black and white will be muddled into shades of grey. In this fog, you will lose the compass we gave you, you will begin to question each of our beliefs; you will grow arrogant like your deceivers, believing you can decide on each truth individually. (As if infallible truths are matters of personal judgment! As if it is possible that we would mislead you!)

The corruption will advance gradually from one issue to the next, ever closer to our core tenets. At some indistinguishable point, you will have so transformed that you will be, in fact, no longer one of us; you will then feel the loneliness of your choice, the loss of the certainty, community, and righteousness that we offered you.

But this does not have to be. You can conduct your life to have

only those associations and experiences that will support and rein-
force our ideas; you can zealously prohibit the corrupting ideas
from being heard; if heard, from being considered; if considered,
from being influential. You can use more skillfully than the
deceivers all arts of pleasantry, reason, rhetoric, skepticism, faith,
politics, and rage to exclude these ideas from your heart and mind.
If you do not do this, if you allow yourself to feel the slightest
inkling of sympathy for the idea, then you are feeling the lovely
sensation of the venom entering your vein.

Once a young person has been trained to judge ideas not by their merit
but by their degree of conformity to preferred conclusions, they are inocu-
lated against change. Thus, one has the assurance that the only new ideas it
shall receive are affirming ideas, the ideas that build upon those that are
already known and never fundamentally questioned. Such is the ultimate
victory of censorship, when the person has been trained to be his or her own
severe and ever present censor.

The worst things are those that are novelties, every novelty is an innovation,
every innovation is an error, and every error leads to Hell-fire.
        —*Attributed to the prophet Muhammad*[16]

The Scholastics are defined by their unique variation of this self-
censorship, which allows alternative ideas a little sport. The Scholastics tolerate and even encourage investigation or consideration of alternative and
even heretical ideas, so long as in the end their conclusions are the same as
established doctrine. Considering themselves fair and open minded, they
engage in discussions, take classes, and read books that they know are contrary to their cherished ideas. They may even indulge in a little pride for
their tolerance, forbearance, and cosmopolitan interests. But from the
beginning, there is a determination to never change one's mind, no matter
what the weight of evidence discovered. Ideas are considered out of curiosity, amusement, or, perhaps, "to better know the enemy." They may earnestly
believe ahead of time that such challenges are valuable because "they will
only make me stronger in my belief." There is neither a genuine unbiased
assessment of the relative strengths and weakness of competing ideas nor a
possibility of changing one's mind. No degree of evidence or argument shall
ever rise to the level necessary to change the mind, for that level is always
reset higher than the evidence before it. Scholastics can explore whatever

ideas they wish, but their final conclusions must be within the bounds of the set orthodoxy. They fancy themselves great adventurers, but, come nightfall, they always return home to sleep in the beds made for them.

Work Aversion

Sometimes we are just too lazy to be rational. Many of the necessary steps to thinking for oneself require time, planning, delayed gratification, and old-fashioned, sore-fanny, achy-back, red-eyed hard work through late evenings and weekends. Yes, some of those times when we give ourselves an extended leave of absence from such labor we are enjoying well-deserved rest and relaxation, but how much relaxation and entertainment is really appropriate? It is good and healthy not to labor fruitlessly or to be less inspired if the pay does not increase with the work, but efficiency and economy matter little if, at the end of the day, the year, or the life, the actual amount of work done was far less than the potential. Thus, we may rationalize ourselves into ignorance, failure, mediocrity, or missed opportunities for greatness. We hear a bitter man grumble at work, "It's a Monday" and celebrate, "Thank God it's Friday" when, with a different lifetime approach to the hard work of study, what we might have heard was his historic speech on the floor of Congress.

Enter two young Americans walking down a sunny road in rural Kenya.

Jill: What this country needs is more aid from the United States: more books, medicine, food, and especially skilled workers to do the building and missionary work.

Scott: Really? I've been reading about how every time free books are available on one street, some local bookseller on the next street goes out of business. Free food does the same to the farmers and free clothes the same to the clothiers.

Jill: What kind of theory is that? The Bible is very clear: if we see a man with no coat, we should give him ours.

Scott: In the spring issue of the *Journal of International Development*, it said the United States gave millions of dollars to Kenya through the 1970s and '80s—even if only to keep the region from getting too cozy with the Soviets. And what effect do you think it had on the per capita GDP? On test scores? On mortality rates?

Jill: I didn't read those reports. Look around, people are happier. They tell us all the time they are glad we are here. If I have moments of doubt, I simply focus on my faith; perhaps I'll understand the mysterious ways later. The commandment to give to the poor didn't have any escape clauses with it.

Scott: As an American teacher in Kenya, you have a nice home in a nice neighborhood, teach in a well-funded school. You have a very low cost of living in this country but are paid U.S. wages. Could your opinion on international aid be influenced by your personal situation?

Jill: I chose my career because of my opinions. But what about you? If you are right, then you have not only wasted two years as a disillusioned Peace Corps volunteer, you can go home knowing you did more harm than good. Is that the way you want to see yourself?

Yet there have always been those who know that not everything depends on fate, that indeed much of human accomplishment is strongly dependent on human effort and all the more on thoughtful, strategic, calculated human effort. There are those who genuinely enjoy the work they do and who find great satisfaction in the work itself, even if the labor is quite arduous mentally and physically and even if the rewards are abstract and long delayed.

> *I have been a very lucky man. I have also noticed that the harder I work, the luckier I get.*
>
> *—Attributed to Thomas Jefferson,*
> *scientist, philosopher, third U.S. president*

Often this is the case with the scholarship necessary to finding solid evidence for our opinions. Alfred Mander reminded us in this book's opening that thinking is skilled work; so, too, is the effort of finding the evidence to think upon. For the average person who struggles to simply survive and to rise above the abject poverty of economy and intellectual stimulation, the mountain of scholarship to be climbed is too distant and daunting for him to even consider tackling. He may even console himself by sneering at those who attempt it. But those industrious persons who find such scholarship daunting but worthwhile may overcome their natural work aversion and make the supreme effort to learn for the sake of learning as well as for the improvement of themselves and society.

I Am the Worst Judge of My Own Fund of Knowledge

A scale can't weight itself. And individuals cannot assess his or her own fund of knowledge, at least not without some type of outside tool—a comprehensive exam, for example. Why is this?

We tend to be too nice to ourselves. Some computers have more data stored within them than others; some minds have more data stored within them as well. Unfortunately, humans don't come with USB ports for direct downloading of data into brains. Bummer. Since it is so difficult to expand one's fund of knowledge, we become eager to judge our existing fund as sufficient, giving us an excuse to stop the laborious learning process. We also convince ourselves that that which we already have—current knowledge, good intentions, strong character, and worldly experience—are adequate substitutes: no more homework.

We can try to assess our actual fund of knowledge, but people cannot comprehend the significance of things they do not yet know. As a result, we err in our own favor: if we know a fair amount about an issue, then we assume we have a sufficient understanding of it; if we know little on an issue, then we may assume there is little to know on the topic that really matters; if we encounter some new ideas, but they are so unfamiliar as to connect poorly to our current knowledge, then those new concepts must be "impractical."

Instead, we may judge our fund of knowledge against other persons in our daily life; but as products of the same culture, they may be doing no better than you. It's no reassurance to be the cutest toad in the pond.

We need a higher standard to compare ourselves to. A test, perhaps. Or ourselves now versus ourselves prior to learning. It is after learning that we realize how poor our knowledge base had been, how relevant and useful the new information is. Thus, we have to keep learning about things despite our natural reluctance to believe that there is anything meaningful left to learn.

If we want to be uncommonly helpful and to give an uncommonly wise answer, we have no choice but to build an uncommonly broad and deep fund of knowledge. This, in turn, requires an uncommon amount of time and effort. There is no way around this; it is a natural law. It is the curse of tabula rasa: each of us was born with a mind similar to a blank slate and it is up to us to put useful

Consider this parable: "I'm Done Climbing." Imagine the speaker as a hiker somewhere on a steep slope of a mountain. He feels as though he's been hiking a long time and he's tired. He has no idea how tall the mountain is. All he knows is that people have compelled him to climb, saying life would be better for him up high than in the valley of ignorance below. There is a vast view behind him and a limited view of a steep rise ahead.

"It is easy for me to believe that, as of today, I already know pretty much all I need to know in order to have a good and fulfilling life. I have, after all, learned much and climbed high on this Mountain of Knowledge. Surely I am near the summit—or sufficiently near it. When I consider what I have recently learned, I am thrilled, and I love learning. I look down on those behind me with sympathetic pity—how uninformed and naïve I was when there! I amaze myself with how far I have come. It was hard, but it was worth it.

"Surely, there is no reason to worry about going any farther and no purpose to continuing the labor anyway—all that seems worth knowing, I know. I have spent practically my entire life getting here, and I'd like to move on to the tasks that give life meaning. No one would blame me for stopping here. Nor will I look with contempt at those who may be above me—no one I have heard of—they who fritter away their lives wrestling with useless esoteric information, studying things that I am sure don't matter. If they mattered, I would know all about them already.

"Sometimes the fog clears enough to glimpse a bit of what lies ahead, but from here, nothing looks very significant, or at least not worth the trouble. Two ideas or groups said to be quite different and in opposition to one another actually look from here to be virtually indistinguishable. Suunis? Shiites? What does it matter? Existentialism? Humanism? What's the fuss? Those people who make so much of those ideas and the differences between them need to get a life.

"However, I admit that, once or twice, I went through the labor of hiking over to those distant ideas and explored them. Okay, I was dragged to them. And yes, they were more interesting than I had expected, and the differences much more significant. But I know that stuff now; the things farther ahead aren't likely to be that interesting or significant. What am I supposed to do? Keep learning my whole life? That's what school's for, and I'm done with school.

"I know more than most people and I know enough to get by in life. I'm healthy, I'm happy—usually—and don't really see a need to keep putting out the effort. At this point, what I do not know and what doesn't come easily isn't worth the effort of learning."

knowledge on it, in breadth and depth. There are many questions that alone can be the labor of a lifetime.

Ah, Misha, he has a stormy spirit. His mind is in bondage. He is haunted by a great, unsolved doubt. He is one of those who don't want millions, but an answer to their questions.

—of Ivan, the intellectual,
The Brothers Karamazov
by Fyodor Dostoyevsky

Is the Idea Inherently Open to Scrutiny?

Few ideas are not open to scrutiny, but there are some. For example, some notions are conveniently beyond the scope of testing: "All the detectable universe is an atom of a greater world." Beyond the limits of what is "detectable," you can place anything you wish, just don't expect people to believe you.

"The god Jupiter governs the universe but prefers to remain undetected, except through faith." Then how can you know anything verifiable about him? How do we separate the hopeful delusions of all the other religions, which are equally certain in their faith, from our faith in Jupiter? Is our community somehow exempt from this otherwise universal mythmaking tendency? How would we know if human fabrication began to pollute our divine doctrine?

The built-in inaccessibility of such issues is different from blanket proscriptions in which the inability to openly scrutinize an idea is by simple decree: "The decision was made (by the CEO, the mother, the leader, and so on) and is final. There will be no further discussion." Such issues will indeed be scrutinized, but out of earshot from the person who fears and thus oppresses the scrutiny.

Scrutinizing an idea is not a matter of whether an authoritarian figure *wants* it to be open to scrutiny; despite threats from an annoyed parent or other intimidating authority, most ideas are inherently open or not open to scrutiny.

The Frankenstein Effect

What was built by people to serve people takes on a life of its own, becoming a powerful entity that demands *we* serve *it* instead. Government bureaucracy is the poster child of this—whether justified or not. What had been constructed to serve the people bulked into a powerful force that is accused of bullying, exhausting, and imposing on the them. But other examples of this type of thinking go far beyond bureaucrats:

- "Implementing any gun control is absolutely wrong, because it violates the Second Amendment of the U.S. Constitution." But which is the tail and which is the dog? The intent of the argument for gun control, of course, is that regardless of what the Second Amendment says, the law can be changed.[17]
- An argument against school vouchers is sometimes heard that sounds rather similar to, "We cannot let our families switch to better schools because that would be bad for our current school system."
- Moral codes are contrived entirely by humans for their health, justice, and welfare, but have been powerfully reinforced by dire threats and promises of infinite reward that, even if the ancient codes were shown to be obsolete or detrimental in some contemporary circumstances to the health, justice, and welfare of the people, obedience to the ancient code would somehow remain the priority.

In every case, we suspect the Frankenstein Effect: we built the government, the Constitution, the school system, and the moral codes, but when powerful evidence arises that there are serious inadequacies and harm in the system and the time comes to make careful and prudent improvements, the system we built demands we serve it rather than it serve us, and we choose to comply.

Forgetting the Opposite Error

When we make judgments, we are supposed to seek out and consider the evidence and ultimately choose. For example, jurors listen to all the evidence presented in court and then decide "guilty" or "not guilty"; voters decide "yes" or "no" to ballot initiatives; leaders decide to "invade" or "not invade." (Many decisions are, of course, not dichotomous, but we'll keep the situation simple for now.) In each of the examples presented here, there are two opposite errors that could be made, such as in the first example: the person might have been found guilty when they were, in fact, not guilty; or, conversely, they were found not guilty when they were, in fact, guilty. Because the information necessary to make a decision is almost never perfect, even in the hands of the wisest, most fair, and objective person, there will almost always be a risk of both types of errors. Over time, and with many such judgments, both types of errors will occur.[18]

Our first frustration is with the inadequacy of our data. Obviously, we want better information on which to base our decisions. And perhaps through technology, or better systems, or human education, we can get better information over time for future cases. But often we find that we must decide *now*, with the information that we do have, and, in the long run, with many such cases, the two types of errors are inevitable.

At the time of the decision, was it an error for George Bush and Congress to order an invasion of Iraq? With the limited information that they had then, what opposite error did they also have to consider? Answer: they feared not invading when, in fact, they should have; that delaying the invasion of Iraq until they had more evidence of a nuclear threat would end with a mushroom cloud over Miami. Faced with making one of these two types of errors, how much evidence does one require before acting? The decision determines which error one fears the most. If, in this example, they had used a higher standard of evidence—that is, if they had expected more evidence and higher quality evidence that Iraq was an imminent threat to the United States—before deciding to invade, then the likelihood of the second type of error would necessarily increase.

When the government gives out welfare, it is difficult to accurately determine who "should" get welfare and who "should not." Thus, there are two types of errors that are always made: some persons qualify for welfare when in a perfect system they would not have and some persons will be refused when in a perfect system they would have rightfully qualified. Suppose we decide that it is more egregious an error to deny welfare to a needy family than to give it to an undeserving one, so we have a relatively low standard for evidence that justifies the enrollment. That would reduce the incidence of the first error but inevitably increase the incidence of the opposite error.

When we increase gun-control requirements and keep guns out of the hands of those who may not be responsible with their use, what error necessarily increases?

If we know there is risk for two types of errors, then we must ask ourselves if one type of error is worse than the other. Perhaps in our civil courts we decide that we have a greater fear of finding an innocent person guilty than we do of finding a guilty person innocent. Thus, we may choose to have a very high standard of evidence for finding guilt and reduce the risk of declaring an innocent person guilty. But, in using such a high standard, we also automatically raise the likelihood of finding a guilty person not guilty. Thus, our second frustration is that we cannot immediately reduce the total amount of error, but can only shift it from one type of error to another, all according to the value we place on each type.

Each type of error has a constituency. Regarding the decision to eliminate Saddam Hussein as a potential source of terrorism, the possible errors were to invade and occupy a nation that was not actually a growing threat to the United States or that the United States would suffer a terrorist attack by Hussein that might have been prevented by an invasion of Iraq. The people of New York City are likely to view the relative weights of the two errors differently than the people of Baghdad. If one constituency's concern is more influential in the decision-making process, then the standard for evidence will be set in their favor. That is exactly what happened.

The third frustration is that better information often comes to light *after* the decision has been made. At that point, we are glad to have more information, we just wish it had been available sooner. It is tempting to pass judgment on the decision and the decision maker after better information is available, but the person can only be held responsible for the information that was available to him or her at the time. (However, he is, of course, also responsible for his earlier decisions that led to the type of information he had available for the present decision.) Finding more evidence after an honest and unprejudiced judgment is useful, historically, and may warrant a reversal of decision after. It may inform us as to how we may have better information for such decisions in the future. But it does not add to the calculation of the competence of that original decision per se. Unprejudiced judgments are, well, judged in light of the information actually in hand at decision-making time.

The fourth frustration is that the decision maker is not always the "wisest, most fair, and objective person." Should a judge who has recently been assaulted preside over an assault case? Perhaps she'll choose to rush a verdict that could have reasonably waited for better information; perhaps she'll allow some types of evidence preferentially; or perhaps she'll misinterpret the evidence by ignoring something that was valid or exaggerating something else beyond its merits. There are reasons enough for competent judges to make errors (such as having incomplete or erroneous information), so to compound them with a prejudiced or corrupt system of gathering and assessing information is devastating to the integrity of the decision-making process.[19]

How then do we manage these two types of errors? When a speaker points out one type of error—the welfare queen, for example, or the apparently reformed felon who's been sitting in jail for decades rather than having been released—we must acknowledge that the opposite type of error exists simultaneously and look for the evidence of how often we make each type of error and if that seems to us the optimal, if imperfect, balance in an imperfect world. We must assess the balance or prejudice with the information that was collected and analyzed and seek to make improvements in the system and the community so that such uncertain decisions will arise less frequently.

Next time you hear someone complain bitterly about the following errors in our society, ask them if they have considered the opposite error. What do you think the opposite error would be in these examples?:

- If we teach comprehensive sex education in high school, some students will complete the course with so much knowledge that they will no longer have a healthy fear of the adverse consequences of teen sex.
- If we allow some cancer patients to relieve their suffering with marijuana, some people without cancer will abuse the system to use marijuana recreationally.
- If we give the government too much freedom to look into the personal lives of citizens, many innocent persons will have their privacy unnecessarily invaded.

The Hierarchy of Data Validity

All information is not created equal, and most of it is neither all good nor all bad. While, indeed, much "information" is wild speculation, false rumor, urban legend, bad science, dogma, propaganda, and outright deceit, and while some is irrefutably substantiated, most information is somewhere in between—it may seem reasonable or plausible *to a degree*. We have an obligation to assign confidence to an idea only in proportion to the validity of the evidence supporting it, no more and no less, even if for other reasons we love the idea or hate it. We need a way to measure the true worth of individual data. Is it complete? Accurate? Trustworthy? And we need a method to weigh them against each other. You are the jeweler of information, and people are thrusting pieces of information at you, each claiming that it is very valuable. You stay aloof, inspect each piece critically, and assign its true value, apart from their claim. When analyzing the relative merits of opposing theories, we cognitively place information into both sides of balancing scales and weigh them against each other.

What you need is a measuring stick against which all data can be assessed a value, good and bad. In fact, one may imagine just such a measuring stick in the form of "The Hierarchy of Data Validity" on page 84. While a jeweler may deal with gold versus fool's gold, diamonds versus common quartz, you deal with firsthand witnesses versus hearsay, evolving understanding versus blind dogma, peer-reviewed statistical studies versus shoddy newspapers.

This is a very rough approximation of relative merit, but the idea of a gradation is still instructive. At each level of the spectrum is a corresponding degree of confidence justified by the nature of the data.

When a person is trying to convince us of something, we should refer to this hierarchy to remind us of how much certainty we associate with such information: Is this mere speculation? Is this an off-the-cuff opinion from a person with minimal experience? Is this person impartial? Or is there an underlying financial interest or ego interest that could be coloring the story he is presenting to me?

We have an obligation to think probabilistically, determining the relative weights of arguments for and against each possibility. This often lands us squarely in the grey zone, where most of us are uncomfortable; how much simpler when issues are black and white. Very seldom does a single piece of information, a single article, a single scientific study carry enough weight to decide a question definitively. Usually there are conflicting bits of evidence, each of which has some merit. What is usually required is comprehensive data, methodically and objectively collected, each part of which is scrutinized for its validity according to explicit standards of evidence, and all considerations are applied to arguments for the various alternatives argued. Even then, an argument seldom achieves 100 percent proof or disproof. But 100 percent proof is not necessarily required to make an informed and reasonable decision. We should make our best decision based on what we have available, but not stop researching, as we are indeed obligated to continue considering new information as it arises so we can change our minds later, as necessary. For now, we should allow degrees of confidence in our decision based on degrees of evidence. We cannot justify claiming 100 percent certainty based on only a limited amount of evidence or justify rejecting a high

Certainty

High confidence

Confidence

Cautious

Speculation

No information available

Opinion deferred

Anecdotes

Question certainty

Reject certainty

Oppose the source

Mathematical certainty

Absolute fact

Well-designed scientific studies: prospective, randomized

Consensus among experts who practice open scrutiny

Scientifically substantiated theory

Retrospective studies

Patterns of history, psychology, sociology, ...

Collective experiences from a large number of events over a long period of time

Personal experiences from a large number of events over a long period of time

Hypothesis

Personal experiences from a small number of events

Fragments of information

Emotional opinion

Poorly designed studies

Advertising, sales, and marketing

Agenda theories: consolation, esteem, fear, hope, ...

Biased "scientific" studies

Half-truth and partial information designed to deceive

Intentional deception; very counterproductive data

degree of confidence when the evidence, objectively considered, tips distinctly one direction.

Note that aspects of the scientific method appear near the top of the hierarchy. Science is an especially successful model for critical analysis skills. Further on we will look at some of those particular skills.

Self-Exemption
Critical comments carry little weight if they seem hypocritical.

- "I'm outraged that the Republicans would be so fiscally irresponsible as to spend more money than we have!" (Memory may be selective sometimes.)
- "I'm shocked—shocked!—that those Democrats are blocking judicial nominees largely for political purposes!"

Remember when Mom told you that whenever you point a finger at someone, three fingers are pointing right back at you? Sometimes adults forget the basics. We do better when we focus on the behavior itself as contemptible, wherever it occurs, and seek to eradicate it every time we see it, regardless of our preferences and allegiances.

What are some of the errors happening in this boardroom?

Tammy: As you all know, the Oversight Committee was formed six months ago by volunteers here in the office to try to do some research and study why the investment firm is struggling. We have put tremendous effort into the studies ...

John: And we've paid you fairly well for your time.

Tammy: That's true, John, thank you. Anyway, we've realized that to ensure better leadership in revenue generation, we need to make some changes, so we have decided to make cuts from client outreach and research staff.

Phyllis: What about cutting the budget of the Oversight Committee?

Tammy: Well, that doesn't make sense does it? The Oversight Committee is our solution to this problem. That will be the one area in which we need to actually channel more of the budget.

John: Any other suggestions?

Tammy: What we need to do is a better job of selecting clients in whom we invest. Right now, we take too many chances on too many investments that do not pay off, so we are going to reduce the number of investments that we make. We also need to remove anyone from an investment research team who has personal stock in that client—we don't want personal self-interest influencing the firm's decisions.

Non Sequitur

"It doesn't follow." A non sequitur is any pretended jump in logic that doesn't work cleanly, perhaps because of unfounded premises, unmentioned complicating factors, or alternative explanations, such as "This war is righteous because we are French!" or "You will do what I say because you are my wife!"

> *"Why the hell didn't you shoot? ... What if the bull [elephant] had charged?"*
> *Farah the faithful produced another drink, and Blix produced a non sequitur. He stared upward into the leaves of the baobab tree and sighed like a poet in love.*
> *"There's an old adage," he said, "translated from the ancient Coptic, that contains all the wisdom of the ages—'Life is life and fun is fun, but it's all so quiet when the goldfish die.'"*
> —Beryl Markham, West with the Night

In literature, non sequiturs can be delicious. In logic, however, they can be unhelpful and misleading. Many of the specific fallacies of logic discussed in this book are special forms of the non sequitur.

Flawed Syllogism

This term may sound like a frightening and expensive problem with your car, but it's just a problem with your reasoning, which can be much easier to fix. A syllogism is the simple device, first described by Aristotle, of making a conclusion based on two premises. For example,

All that store's bread is made from wheat. This loaf is from that store. Therefore, this loaf is made from wheat.

Pretty simple, huh? You'd be surprised how many ways we can screw it up. Consider first that there are at least four components to a syllogism: at least two premises, a conclusion, and the logic (supposedly) used to connect them. The syllogism is flawed when one or more of these things occur: one of the premises is wrong, the conclusion not valid, or the logic unsound. Often one flaw causes more flaws. Can you identify the flaw(s) in these?

1. All birds can fly. Penguins are birds. Therefore, penguins can fly.
2. I want a good grade in this class. I always get what I want. Therefore, I will get a good grade in this class.
3. The biscuits came out of the oven. The cat gave birth in the oven. Therefore,

the cat delivered a litter of biscuits.

4. All flowers are plants. All daisies are plants. Therefore, all daisies are flowers.

The above examples serve to demonstrate some of the most important tools and concepts in the analysis of an argument, which are:

- Truth: whether any given statement concords with reality
- Validity: whether the reasoning is correct, apart from the conclusion being true. In colloquial use, validity may also refer to the conclusion itself—being both true and the product of valid reasoning.
- Soundness: refers to the argument as a whole. When the premises are true and the reasoning valid, the conclusion is true.

As to our examples above:

1. One premise is not true, so although the reasoning is valid, the conclusion drawn is not true.
2. The same as 1.
3. Both premises are true, but the reasoning is not valid, so the conclusion is not valid. The conclusion also happens to not be true.
4. Both premises are true and the conclusion is true, so we may miss the fact that the logic was not valid.

Of course, people don't usually speak in well–laid out triads of premises followed by conclusion. Most often, people don't mention the premise that they consider well understood. Consider the examples below.

- The skies are clear. I don't need my umbrella.
  Translated: I don't need my umbrella when it doesn't rain. When the skies are clear, it doesn't rain. The skies are clear. Therefore, I don't need my umbrella.
- You will do what I say because you are my wife!
  Translated: You are my wife. All wives must do what husbands say. Therefore, you will do what I say.

In this last example, he left one premise unspoken, and it wasn't the one that was the most self-evident. Perhaps he wanted to sound logical but sensed that his other premise was vulnerable on the point of its truth. An essential skill in the analysis of an argument is identifying the unspoken premises and verifying their truthfulness.

Many students try to master the semantics of when a conclusion is valid but unsound or true but not valid, and so on. Although Aristotle might roll in his grave to hear me say it, I do not think mastering those terms is the critical issue; we have bigger fish to fry. If you can consistently identify that there is a syllogism in the discussion, what its components are, and where the flaw is, then you grasp the main point.

- Translated: Many students try to master the semantics. This fine point is not the critical issue, relative to our present goal. We need not master the issues that are not critical. Therefore, we need not master the semantics.

In this chapter, we have explored some of the fundamental errors that most people tend to make, whether in everyday discussions or in historic political debates. Tools for detecting these errors, for identifying them, and for refusing to accept them were given. These general concepts serve as a framework for our study of other, more specific lessons on rational thought that we derive from history, science, religion, rhetoric, human nature, and mathematics.

There is a realm of human experience that has a long history of grappling with and testing rational and irrational ideas. It has struggled imperfectly but very productively against babble, bias, and hype for centuries. From it, there is much to learn. Let us look now at the lessons from science.

Chapter Three
The Scientific Method in Everyday Thought

> *If we confine ourselves to exhibit the advantages deduced from the sciences in their immediate use or application to the arts, whether for the welfare of individuals or the prosperity of nations, we shall have shown only a small part of the benefits they afford. The most important perhaps is, that prejudice has been destroyed, and the human understanding in some sort rectified; after having been forced into a wrong direction by absurd objects of belief, transmitted from generation to generation, taught at the misjudging period of infancy, and enforced with the terrors of superstition and the dread of tyranny.*
>
> —*Marquis de Condorcet*, Sketch for a Historical
> Picture of the Human Mind, *1794*

Learning How the World Works
and Using Rational Thought Are Not Separable

Almost everyone is interested, to some degree, in understanding how the world around them works and understanding just what are the various forces, seen and unseen, that affect and influence their lives. Since Newton demonstrated the existence of natural universal laws, humanity has known that these laws can be studied, understood, relied upon, predicted, and even harnessed to our own purposes. And because those machinations can be complex, their deconstruction requires rational thought. The method that rationally examines phenomena we call "science," a word derived from the Latin *scire*, "to know." Thus, understanding objective reality and science are not separable; they are two sides of the same coin. For persons interested in staying alive, breathing is not an option; for persons interested in having valid objective knowledge, science is not an option.

But not everyone is interested in having valid objective knowledge. It is certainly possible to use personal computers, fly on airplanes, and take lifesaving medical therapies with the same obliviousness as we may have when we use our kidneys and the oxygen around us. We do not have to bother to consider the origins of such things—or we may accept easy, if absurd, answers for our occasional questions—and go about a very uninformed, misinformed, and contented life. It is even possible to devote one part of one's life, even one's career, to science, and yet keep such thought so well compartmentalized that when we turn off the lights at the lab to go home, we resume our mystical and uncritical thinking in all other aspects of our lives. Sir Isaac Newton himself is infamous for his bizarre and mystical beliefs outside of his world-changing discoveries in physics and mathematics. Reason is always optional, of course—it can be turned on and off as that light switch—and many people choose to opt out and not be rational or to be rational only when it suits them. But if the person opts out of rational thought, then he or she is opting out of having a higher and more objective understanding of how the world actually works in that particular aspect of his or her life.

Science is an imperfect tool; the history of science is one not only of astonishing discovery and contribution, but also of innumerable errors, famous foibles, and dire disasters. Science gives us vaccines, clean water, seat belts, moonwalks, genetic decoding, quantum physics, and deep-space astronomy; it also gives us thermonuclear warheads and genetically engineered biological warfare agents. It may be a tool that can be used for good or for ill, but we know this—it works. Rational thought is powerful, delivering results far more reliably than wishful thinking. Every time we hold a book composed of manufactured paper and ink, we are benefiting from rational, concrete, undeluded thought—even if that book promotes mysticism and the occult. The store that sells crystals and magnets to "heal" your body is built of and well lit by rigorously scrutinized scientific processes. Even when our high-tech devices break down, science can lead us to the explanation and to the repair. Science may take some wrong turns and go into dead ends at times, but, for objective matters, over the long run, done correctly, it delivers. Thus, science is a valuable source for ideas on how to think clearly and logically in many other fields, including business, public policy, and relationships. The following are some of the fundamental principles of the scientific method that we can apply to everyday thought.

Nothing Is Beyond Question

There are no premises or assumptions that are not fair game for scrutiny, reassessment, and infinite reconsideration. No ideas, no persons, no authorities, no institutions, no policies, no myths are exempt. When scientists are told their method does not apply or that they should not try but should leave it and move right along, they smell humbug and are attracted all the more. Scientific *thought* is allowed to *start* and proceed with anything. Scientific *practice* can almost always follow with important limitations due to ethical considerations—and even those ethical or moral codes are open to rational scrutiny. Science is incomplete when it has failed to scrutinize anything; it is unhealthy when it has chosen not to scrutinize something for inappropriate reasons, such as the preservation of cherished error and misleading myth.

There Is No Preset Range of Acceptable Answers and Unacceptable Answers

Scholastics allowed students and teachers to consider any question and exercise any thoughts, so long as the final conclusion accorded acceptably with the given orthodoxy of the times. This is stillborn science: conclusions must be only those that are reasonably and objectively supported by the evidence, no more, no less. Reason is allowed to *conclude* with anything that the evidence supports.

Scientists admit that they are frequently wrong. Individually and collectively, their very goal is to identify the errors of our understanding and correct them, without end, creating an ever changing and ever improving body of knowledge; this is a celebrated and admired accomplishment in the scientific community. Is that the case in ourselves? Our families? Our jobs? Our public policy? Persons expecting certainty, absolutism, infallibility, or permanence will need to look somewhere other than science.

Blind Faith in Science Is Appalling to Scientists

- "No real scientist would question that idea."
- "They wouldn't have printed it in an official journal if it weren't true."
- "They wouldn't have let this drug be on the open market if it weren't safe."

Determining the degree of validity and applicability of a scientific study is very hard cognitive work and requires specific skills of a particular methodology. Uncritical trust in scientific sources is dangerous and not in keeping with science. Rather, it reflects naïveté or fatigue or, regrettably, laziness on the part of the reader. Scientists, being normal, fallible persons, can be guilty of those things too, and, despite good intention, they may produce poor and misleading research: they may be funded by a biased source that publishes only the answers they like; they may simply make mistakes; or they may miss

Read each article you come across with a skill-fully critical eye. Ask yourself:

Is this original research or do they just make claims based on other "studies"?

Is it the original research itself or just a "glossary brochure" that summarizes it?

Is the tone neutral or does it use emotional words such as "insidious," "astounding," "breakthrough"?

Does their question assess all possible and reasonable answers?

Did the study control for all other reasonable variables?

Was the sample size adequate?

Was the method of data collection unbiased?

Did the researchers or publishers have a financial or personal interest in the outcome?

Did the design allow for double-blind analysis?

Is enough explanation given that you or others could precisely repeat the study?

Does the author voluntarily point out and thoroughly discuss the study's weaknesses?

Do the conclusions overreach the context of the study?

Are the conditions of the study similar enough to the conditions of my situation to reasonably apply its conclusions?

the flaws in the study that greatly reduce the validity of the conclusions. A great deal of "official studies" and "academic research" are poorly designed, meaning that the conclusions are not adequately supported by the data presented. In the worst instances, the scientists slide from poorly performed research to intentional deceit: they have an explicit and hidden bias, in which case their work is, by definition, no longer science; they may abandon scientific principles and choose to knowingly conduct seriously flawed studies or to misrepresent their findings. Such practice is not worthy of the name science. Exposing charlatans is an important responsibility of scientists, whether the charlatan is operating inside or outside the scientific community.[1]

However, real scientists doing real science are defined by their integrity to the objective method. They make no claim to ultimate and infallible truth. Instead, they point out the weaknesses in their studies; they consider every conclusion a temporary one until better science gives a better answer; and they constantly change their opinions as better data indicates it. Legitimate scientific studies must be carefully designed for validity, and even after publication, they must be carefully scrutinized for flaws that undermine its validity or applicability.

Everyone, inside or outside of science, can learn from that type of objective and logical rigor. We can resist the temptation to claim infallibility or to bury the weaknesses or uncertainties of our information. We can hold our conclusions as tentative and change them despite the test to our pride.

Emotionalism Is Discouraged

Scientists are human, and they get attached to their pet ideas. But, similar to farmers with their cute little pigs, they know they shouldn't get attached because they know that most of their ideas are going to come to their natural end in the market of scientific scrutiny. Scientists know that their task is to produce and develop a robust idea, to deliver it up to the public, and to allow the public to take it apart, study it, ingest it, and benefit from it, all to the betterment of society. It's hard to watch that happen to your pet idea. Scientists know they need to refrain from being emotionally attached to their opinions and beliefs, and so do all of us, but since that is hard to do, we, too, have to allow public scrutiny and be willing to accept the verdict of more objective assessments.

In everyday life, as in science, we must ask ourselves *why* we believe something. Does the idea give you some form of gratification that is important to you? Was it because you were told in your youth that *this idea* or *conformity* was critical to being considered a good person and a person who deserved respect and admiration and praise? Do you identify the idea with your family or with friends whom you love? Do you feel that to reject the idea would seem to them (or to you) as if you have rejected the family? Have you been explicitly manipulated this way? Have you entangled the issue with your personal and social identity, so it is hard to change? Were you ever shunned by people you liked because they felt you didn't live up to their expectations of adherence to certain ideas? Do you feel as though the people you love, such as parents, would be quietly "disappointed" in you and be less proud of you if you were to express any doubt or contrary argument about some of their core convictions? How much of your zeal for the idea, or even your failure to explore your uncertainties about it, is actually an attempt to avoid creating distance between you and those you love or respect? Have you and the people you like agreed that your relationship is not based on whether or not your opinions are the same?

Investors, too, for example, know well these hazards—investments are made on information that is good or not, but, once committed, an investor who is not thinking rationally tends to remember only information that justifies the investment, forgetting that which argues against it. They then hold the investment long after its value disappoints and long after a more clear-minded person abandoned the turkey.

Substantive Debate of the Evidence Is Encouraged

Scientific meetings are characterized by frequent interruptions with questions from the audience, typically questions that challenge or express skepticism of what is being presented. Scientists usually find that the benefit of the public exchange outweighs any loss of decorum. People looking for truth say, "Bring 'em in!" People looking for a shouting match, or worse, say, "Bring 'em on!" People looking for power say, "Burn 'em at the stake!" It is the scientists who invite criticism and skepticism, and it is a lesson for general society.

To those who may say that science is "just another religion," my reply would begin with the preceding five sections: science is not another religion because in science:

- Nothing is beyond question
- Conclusions are not limited to a preset range of answers
- Blind faith is appalling
- Emotionalism is discouraged
- Substantive debate of the evidence by knowledgeable proponents of all points of view is encouraged, invited, and given top priority at gatherings

I would also cite the section "Believe Now, Understand Later" in the last chapter as another description of a characteristic distinction between religion and science.

Hypothesis Is Not Knowledge

Allow me to illustrate this concept with an example. I have encountered professional and educated persons who firmly believe the following: among people they had observed in their lives, the children of women who gave birth at a later age, say, thirty-five, seemed to grow into sophisticated and well-adjusted adults, but the children of women who gave birth as teenagers seemed to struggle more with behavioral problems and emotional adjustment. They concluded that something about a woman's cumulative life experience can directly and physically alter the eggs within a woman's ovaries, conferring related benefits to the fertilized egg and thus to the child itself. In fact, the idea is so appealing to them that they have decided outright that this is, in fact, true, and they believe it deeply.

Amazingly, this has happened to me with unassociated people on separate occasions just this year. Our conversations went something similar to this:

"Well, that's a very intriguing idea," I said. "It's a hypothesis, and it's interesting to think how we would test the hypothesis. First, we would have to study big populations of women to see if, in fact, those born to older women fared better as you suspect. Then we would have to try to separate out the difference between nurture and nature, that is, the way an older woman raises children versus the way a teen raises children could explain much of that difference. If nurture did not seem to explain the difference, then perhaps we could examine the eggs from sisters of greatly different ages and see in what ways those eggs had physically changed with time so we'd have a plausible biological explana—"

At this point, my acquaintance was not only bored but obviously annoyed. "It's not a science thing. You can't approach this as a scientist because you cannot uncover the workings going on here. I am talking about the life experiences of the woman having an effect on the *essence* of the egg itself, which produces an inherently more sophisticated embryo. ... "

"What exactly is the *essence*? What is it made of? DNA? Mitochondria? Cytoplasm?"
　　"No! You are such a scientist! It can't be detected physically. It is the spiritual component of persons, it's *energy*."
　　"Hmmmm. Energy is measurable. ... "
　　"Not this kind of energy."

My acquaintances are very intelligent, have degrees from respectable universities, and have come up with this very interesting hypothesis. They were trying to discover and understand the workings of the world, but they made a series of errors that will prevent their idea from being accepted by a scientific audience: they decided the idea was obviously true without submitting it first to valid testing (indeed, any testing at all). That is, they leapt straight from hypothesis to conviction, mistakenly thinking that a belief in understanding how the world works and valid science were separable. They did not concern themselves with the small numbers of women involved in their initial observations; had not considered other possible explanations for the effects they thought they saw; were unperturbed that, for their theory to work, they had to suspend or rewrite natural physical law; were instantly certain the "essence" and "energy" existed, even though they were not detectable; and held firm to their belief that this idea was "above" scrutiny and reason. They were so attached to the idea that they became annoyed that it would be called into question.
　　In order to discover a truth about how the world works, a hypothesis is

a beginning, not an end. Statistically, an armchair hypothesis has a short life expectancy in the no-nonsense arena of science, and a hypothesis sent up primarily for its emotional appeal is likely to come to an end as inglorious as it is swift.

> We are certainly not to relinquish the evidence of experiments for the sake of dreams and vain fictions of our own devising.
> —Sir Isaac Newton (1642–1727), professor of mathematics, Cambridge University, Principia, 1687

Welcome to the Hamilton High School faculty meeting.

Jim: The problem here is that we need to improve student test scores.

Patrick: What's holding us back?

Jim: Well, there are several factors, but some of them are passing students up from the previous grade who are not competent to move up, uninvolved parents, overworked teachers, behavior problems with students, ...

Martha: Okay, let's hold back every student who doesn't pass a standardized test, have parents sit in at least one day each semester on each class that a student takes, expel problem kids, ...

Jim: What? That's ridiculous! We can't do that!

Martha: Why not?

Jim: Because it would probably be ineffective, expensive, and contrary to our basic principles to leave no child behind. No way. Other ideas?

Martha: "Probably"? Why don't we try it?

Connie: Because other counties have tried it and I heard it didn't work.

Martha: You heard it didn't work. Do you have any kind of report or data on their experience?

Connie: No. But the system as it is can work out its problems. You just have to believe in the folks running the schools. They are good people.

Martha: Good people, but the schools have gone from bad to worse. They aren't delivering.

Selective Selection

Don't let your current opinions screen out opposing ideas. We tend to feel an aversion to information that seems, at a glance, to disagree with our preferred notions. If I am a political conservative, I probably don't subscribe to liberal magazines, and vice versa. When we pick up our mail at the end of a busy day, we don't care to be challenged. We enjoy reaching into the mailbox and pulling out reassurance, vindication, unanimity. We may pick our friends, our books, our radio stations, and so on, in the same way. So, consciously or subconsciously, we self-censor, we limit our own exposure to new ideas, but we don't learn much that way. Many of us need to actively search out more diverse and challenging input.

In an age in which media, advertisers, tabloids, and enthusiasts bombard us with input, we quickly realize that the problem isn't one of an inadequate volume of information, but an inadequate *quality* of information. Rather than passively allowing marketers to determine the nature of the information you see, manage it yourself in order to obtain maximum quality: turn off the TV; seek books that give you a comprehensive background on an issue; subscribe to periodicals in which the content is substantial, fairly presented, and contrary to your own opinion; and surround yourself with people chosen for the *quality* of their argument, not for their final opinions. During discussions with others and while reading material on your own, insist that the data is valid, that the sources and perspectives are broad, and that neither you nor anyone else filtered the information in a biased fashion before you began to review it.

Connie: You need to give them more time. You need to trust that they can do it.

Bob: My mother and father both went to this school and their parents before them! We can't go changing the fundamental philosophy of this school! When I look at that letter on my jacket, I think of a school that has stuck by its traditions! (He stands, puts his hat over his heart, and begins to bellow the Fighting Hamsters school song.)

Richard: I think the solution is obvious: organized study groups. Make every student accountable for the grades of three other kids in his class. For every test, each kid gets not his own grade, but the average of the three. That would do it, I'm sure.

Jim: Do you know of any other school that has ever done such a thing?

Richard: No. But it is obvious that the self-interest would generate a lot of social pressure to work together and to pull everyone up. That's the solution.

Selective Observation

Unless we consciously guard against it, our expectations and desires can influence our perceptions. Consider the following examples.

Stephen Hamm, a Dutch naturalist of the seventeenth century, breathlessly reported this on viewing sperm with a microscope:

> I [saw] in the human sperm, two naked thighs, the legs, the breast, both arms, etc. ... the skin being pulled up somewhat higher over the head like a cap.
> —Christopher Cerf and Victor Navasky,
> The Experts Speak

Many mainstream nineteenth-century "scientists" were self-convinced that each ethnic group had a characteristic cranial capacity and contours and that this in turn correlated to intelligence. They even gave their "science" a convincingly sciencelike name: phrenology. Phrenology corresponded nicely with the political agenda of their times, but was devoid of validity. The scientists "saw" what they wanted to see.

In psychology tests, subjects were shown split-second images of standard playing cards and asked to report what they saw each time. When an anomalous card, such as a *red* two of clubs, would be snuck in, the subjects consistently reported "seeing" an expected card—in this case, a red two of hearts. Sometimes in public policy, a perfectly reasonable proposal is made, but because it was made by a member of an opposing political party, we may "see" only all the possible weaknesses and none of the possible merits.

When a sniper began shooting civilians at random, one every few days, in the Washington, D.C., area in 2002, an early "witness" reported seeing a suspicious white van. Soon every shooting had witnesses describing in detail this suspicious white van. But when the case was solved, no white van had ever been involved. By the power of suggestion alone, sincere and well-intentioned people "saw" what did not exist.

Despite Percival Lowell's great contributions to astronomy in the early twentieth century, he may be best remembered for his elaborate drawings of the "canals" that he saw crisscrossing Mars and for his calculations showing that the temperature on Mars was hospitably similar to that of the south of England. Of course, we know now that almost all the "canals" he saw do not exist in any form, and even the ones that may have an objective correlate are part of the natural relief of Mars's terrain and not associated with the artificially built structure that the word "canal" suggests. We also know that the average daily temperature on Mars does not exceed minus 27° F, and temperature variations of 180° F are common—hardly the south of England.

In all of these cases, people, including scientists, "saw" what they wanted to see and what they expected to see. Their observations were, true to the origin of the word itself, prejudiced.

Do you know what you expect to see? Do you know what your biases are? Are you able to admit them? What do you "see" when someone presents an idea concerning abortion, gun control, or a personal ethical lapse in a politician you like versus one you do not like? What do you "see" as the cause when a surprising and favorable coincidence suddenly occurs to your advantage? It almost seems we should hold conclusions that support our expectations and hopes to an even higher standard of scrutiny than usual, if only to try to offset this human tendency to see what we want to see.

Claudius Ptolemaeus was a second-century astronomer in Alexandria, Egypt, whose description of an Earth-centered universe was received better than his predecessors' calculations of a sun-centered "solar system."

Ptolemy, as he is usually referred to, had to invent an elaborate system of "crystalline spheres," each rotating on a connection to another, to preserve the appearance of the motion of the planets and stars.

Fifteen centuries later, Johannes Kepler struggled for a lifetime to reconcile his absolute faith in Ptolemy's crystalline spheres with his own astronomical measurements. We know Kepler's name today because, after many years of fruitless attempts to reconcile the theories, he was willing to admit that the crystalline spheres do not explain the motions of planets, no matter how personally aggrandizing, aesthetically pleasing, and historically hallowed they may be. Carl Sagan describes it thus in his *Cosmos* video series:

> *When he found that his long-cherished beliefs did not agree with the most precise observations, he accepted the uncomfortable facts. He preferred the hard truth to his dearest illusions. That is the heart of science.*

And, apart from science, I might add, it is also the heart of all forms of rationality—at home, work, school, in the legislature, the voting booth, or anyplace else.

Anecdotal "Evidence"

"Did you hear about that woman who was beat up and thrown in the trash by the police? We have a serious problem with police brutality in this town!"

"I read about a guy who got beamed up on a spaceship! UFOs are real, and we are being exploited!"

An individual case report alone often turns out not to be real, and when it is real, it alone is not proof of a pattern, any more than a single point creates a complex picture or even a line.

"I have a cousin who was adopted, and he is kind of strange. I don't think adoption is a good idea."

"It's been a warm summer for New York, so let's go now to Candy Wide-Eye for our story on global warming. ... "

"How is it that I know I can read minds? Personal experience."

An individual anecdote—especially a surprising or alarming one—must be regarded with skepticism until it is analytically verified. Verification requires passage through a series of increasingly rigorous and expensive tests. First, the anecdotal report itself must be verified. If the one case is verified, it may reasonably warrant a systematic search for a pattern of such verifiable cases. If such a pattern is found, then tentative hypotheses can be formulated. The process quickly escalates into a series of ever more elaborate but powerful studies and a tangle of new vocabulary words: a retrospective case-control study that still supports the hypothesis would get serious attention, particularly if it is a prospective, randomized, double-blind, controlled study, which is the gold standard, when possible. These are performed in a methodical way, using sufficiently large numbers to hold up to statistical analysis. Finally, the results of several such studies would be gathered into a meta-analysis, subject to independent review by relevant authorities, and might produce very compelling evidence. We will get more specific with some of these terms later, but the point for now is that there is a rigorous methodology to be applied rather than simply relying on an anecdote.

A single anecdote, or even several anecdotes, does not itself constitute compelling evidence. We may wish them to be sufficient evidence; we may get deep satisfaction from a hypothesis that ascribes the events to a supernatural force, a malevolent plot, an alarming conspiracy, or a comforting psychic connection; but without the proper collection of evidence, such conclusions remain only speculation, which may be fully loaded with alarmist or wishful thinking. Nonetheless, many people can't help but to use a single anecdote, especially a dramatic one, to awe themselves, to dupe an unsophisticated audience, or to excite entire populations into action. Usually the person knows that drier, valid analysis would not support the same conclusion or inspire the same action. But anecdotes can make for both dramatic stories and disastrous decisions.

The Unrepresentative Sample
(Also known as the tortured counterexample.) People are complicated, there are many ideas, actions, and statements made by them or done to them. They are subject to inconsistency; they change over time; they use language clumsily even when they think they are being clear. Picking out one particular quote or action and suggesting it represents the whole may be inaccurate and unjust. A few bad storms locally are not in themselves proof of global warming; a few losses does not yet mean the coach needs to be replaced; a carelessly worded impromptu statement should not be the end of an otherwise

meritorious political career. The complex nature of things can allow persons who are unscrupulous—or who have an agenda, or who have an axe to grind, or who simply have a limited capacity to handle the complexity of the whole picture—to seize upon a single quote or action to pass judgment on the whole.

> "This book is irresponsible rubbish! Look at chapter twenty-nine, page 598, paragraph three, line seven, where it says, 'Indonesia is composed of a collection of small island communities.' I say to you that Borneo is nearly the size of Texas! Small indeed! This book is a travesty of academics. ... "

We do better to remain aloof to the dramatic snippets of information waved wildly in our faces and instead seek out more circumspect information on the entire person or institution in question.

The Meaningless Question

"What happens when an irresistible force meets an unmovable object?" Of course, if there were any such thing as an irresistible force, there could be no such thing as an immovable object, and vice versa. "Governor Frank Bowman, what would you do if you were a seventeen-year-old pregnant girl?" Science and rationality confine themselves to questions that lend themselves to meaningful analysis.

Does the Observer Change the Observed?

Classic errors of this are seen in anthropology when the observed subjects are suddenly on their best behavior or in chemistry when the chemicals react with the equipment rather than with themselves. Surveys are notorious for intentionally or unintentionally soliciting the answers that the surveyors or those who are surveyed prefer. We have to use careful methodology to ensure that the process of our inquiry does not itself distort the accuracy of the data.

Quantify

Measure it in any way possible. This facilitates reproducibility, description, validation, and prediction. A watershed event in the history of the development of science was the establishment of standard units of measure (millimeters of mercury for temperature or meters per second for speed, for example) so that separate people could compare observations and outcomes. Even today, sciences that struggle to make progress are often focused on the issues for which adequate quantification is the most difficult, such as memory, climate, and behavior. Be careful, however, of assuming that your measurement accurately reflects that which you wish to measure. For example, the history of attempts to measure human intelligence is rife with invalid measuring tools and disastrous injustice resulting from too much faith in those tools (see *The Mismeasure of Man* by Stephen Jay Gould).

Selective Differential Explanations

The differential diagnosis, or the "differential," is, in parlance of physicians, the complete list of all possible explanations for a problem. Thus, our first source of bias is in not putting on the list any of the possibilities we do not want to work through or that we do not want to conclude with. But to be objective and rational, we must think of all the different ways that the issue could be explained, including those that are distasteful. Then we must think of tests to systematically support or disprove each alternative, until we are left with one explanation that is significantly more likely than the others.

The old tricks of "horses that can count" and of people "walking on a bed of coals," the surprising incidents in our lives that are surely miracles, and other spectacles are easily explained as very common deceptions and misunderstandings, but only if people allow themselves to consider all possible explanations for a particular phenomenon and then scrutinize each rationally. Does the person allow for any other explanation but the pleasantly fantastic and mysterious?

People do this quite frequently, in matters mundane and profound. Typically, their skeptical friends do not challenge them but leave them to their own cherished though unsupported beliefs. But a true friend may also, in a friendly but honest way, perspicaciously inquire into their friends' beliefs. It is easy to imagine such a scenario:

"Jim, I'm sorry to hear about that fire at your house! My goodness, I'm so glad no one was hurt."

"Well, Steve, everything happens for a reason."

"Have the investigators figured out what was the reason? I've heard that a third of all house fires are due to squirrels chewing through a live wire somewhere."

"That's not what I mean. I mean, there is a higher purpose behind all things that happen, good and bad, so we accept that and move on."

"Hmm. I'd rather find out what the real-world reason was so the same mistake isn't made again."

"Are you suggesting the fire that destroyed my home and all my cherished belongings was my own fault?"

"I don't mean to upset you, but right now, no one knows if it was anyone's fault or just bad luck—such as getting hit by lightning. Uh, you did have a lightning rod on your house didn't you?"

Jim glared. "Even if I had no lightning rod and even if I ignored the signs of squirrels in the attic, even then, the consequences were part of a higher purpose."

"Jim, why do you think that there is a higher purpose? What is your evidence for that?"

"Because I can feel the presence of Shamash. He protects me and guides me."

"Where is Shamash?"

"Everywhere."

"Is he right here with us now?"

"Sure."

"I don't see him."

"There are many real and powerful things we cannot see—gravity, for one."

"But they are detectable in some way, or we could not know anything about them. If I were to spray some green spray paint in the air all over this area, for example, would Shamash then be revealed when I happened upon him?"

"He's not corporeal except when he chooses to be. He's not like some invisible man from science fiction stories. That's silly. Your paint cannot land on him. Perhaps it is like love: love is real and you can't weigh it on a scale."

"Love is an abstract concept. It does not exist outside of the mind. You are making claim to an objective entity that exists in a real way. Can I detect Shamash in any reliable, methodological way? Can I use thermal-imaging devices? Subsonic microphones?"

"No."

"If he is not corporeal, then how can he have a will? A 'will' is a thought, and thoughts are neurotransmitters leaping between synapses of the brain. No physical synapses, no thoughts."

"Why do you suppose that there can be no thought without nerve cells? Doesn't the soul inhabit the physical mind and body?"

"Actually, no. There is no evidence for a soul if a soul is considered some ghostly inhabitant of our bodies. There has never been any evidence in the history of the world of thoughts existing without nerve cells to create them; only scientifically unsubstantiated stories, and for stories there is no end to human creativity. The soul seems to be an ancient and primitive explanation for the thoughts and actions we now can attribute entirely to physiology. I still want to get some kind of evidence for Shamash that cannot be explained by more plausible real-world mechanisms. If it protects you then perhaps I could rush upon you with a knife and Shamash would stay my hand?"

"He probably would not. Shamash does not submit to testing. If you were to do that, then that too would be his will, or the will of evil."

"Well, fine protector you have there."

"He does not necessarily protect me from bodily harm or death. He protects me from anything that is not his will."

"Well, forgive me if I leave Shamash to his own inscrutable will and continue to rid my attic of squirrels."

In the great scheme of cosmic surprises, can we conclude that anything is impossible? While no idea is unmentionable, some, having been mentioned and evaluated, do belong squarely in the realms of fiction or forgivable delusion. Such is the case here, as the theory requires new physical laws or a supernatural suspension of known physical laws, and never once in verifiable history has any object or organism defied physical law, and nothing is offered this time that would persuade a rational observer that this was the first. Jim may claim to have an invisible protector, but there is no evidence to reasonably support it. In addition, when exacting and appropriately designed attempts have been made that would have detected such a thing, if indeed it did exist, and no evidence is there found, then the very lack of evidence becomes itself evidence for the unlikelihood of its existence. Until some or any valid evidence to support the fantastic claim is obtained, all evidence points against it, and the belief in it must be considered irrational.

More interesting, perhaps, is the problem as to why Jim does not see, or refuses to see, that there is no Shamash. For this, there is another differential. You may speculate that Jim was lying, had been told since his youth that Shamash was there and had never scrutinized the idea carefully himself, or had a need for this belief that was so strong as to cloud his own objectivity on the matter. Maybe his parents believed in it; maybe he thinks it is good and noble to believe in it as well; maybe it gives him a sense of safety, comfort, or justice. But having reasons for wanting to believe is different from having valid reasons that support the actual existence of it.

The first breakdown in rationality here was Jim's failure to make a complete differential concerning the causes of the fire. The second was in his failure to scrutinize the possible explanations, including the explanations of his own psychological approach to the issue. The third was in claiming knowledge about an entity undetectable by any verifiable means.

When fantastic events are said to occur, the participants, the audience, and the third parties who believe in and perpetuate the tales often mean no harm in their role. In E. R. Dodds's assessment of the visitations and divine messages at the Oracle of Delphi, he observed, "Anyone familiar with the history of modern spiritualism will realize what an amazing amount of virtual cheating can be done in perfectly good faith by convinced believers."

Occam's Razor

William Occam (or Ockham) was a fourteenth-century philosopher and logician who made the point that when constructing an explanation for a phenomenon, we should avoid fabricating additional complexities without valid evidence for them just for the sake of making our theory work. In other words, if two hypotheses explain the data equally well, but one includes additional complexities for which there is not valid evidence, then the simpler theory without those injected complexities is more likely to be correct, and we will do well to lean toward it until more data is available.

> Example: Your wallet is not where you thought you left it. Have you forgotten where you actually left it or has a burglar entered and carried it away? There is no broken door or muddy footprints on the windowsill. For the latter theory to hold, one must introduce an additional entity for which there is no evidence. Occam's Razor predicts you lost it all by yourself.

Sometimes we do not fabricate an additional complexity for which there is no evidence, but we do the next best thing: we choose to believe a much less *plausible* thing over a much more plausible thing because of personal preference.

> Example: A young man enters a medical office complaining of a headache and abdominal pain and is found to have a fast pulse.

The new intern, eager to make an exotic diagnosis, has recently read that such symptoms could be caused by pheochromocytoma, a rather rare tumor. To be the physician who makes the diagnosis of an exotic and exciting illness is a thrill in the profession that dedicated bird-watchers would understand. Eager for the envy of her colleagues, she orders a CT scan and is sure that a vague shadow near the kidney is a tumor; she convinces herself that the cause of this patient's symptoms is pheochromocytoma. Her attending physician knows that, for all the available data thus far, she could be right, but that the shadow near the kidney is the same shadow that any normal kidney is likely to have. By using Occam's Razor, he instead calls it a routine "stomach flu" with dehydration. Further testing and a little time prove that the attending physician is correct. He smiles at her and says, "When you hear the sound of hoofbeats, think horses, not zebras."

Occam's Razor helps us identify which theories are supported by evidence in all their particulars and which ones require one or more fabrications to make them work.

Additional complexities are often flagrantly inserted to bridge a large gap between the legitimate evidence and the cherished belief. People devise new laws of nature or claim that there is a supernatural being involved or that this idea would be beyond doubt if not for it being suppressed by powerful but hidden conspirators. Issues are often more complex than we first believe, but each of those complexities is entered into the explanation only when substantive evidence for it exists. How much evidence is necessary? That brings us to our next useful tool in critical analysis. ...

Obligation of the Underdog

Carl Sagan is remembered for many important concepts in critical analysis, and among them is his assertion that "extraordinary claims require extraordinary evidence." His perspective was that of a scientist, so the challenge was to all crystal-healing, Bigfoot-glimpsing, alien-romancing enthusiasts to come and prove the scientists wrong. He seemed to have built a fortress of evidence around scientific conclusions and put the burden of proof on the outsiders. Comfy.

The problem is that "extraordinary" is in the eye of the beholder. In some communities, geographic or cultural, the "ordinary" claim is the one scientists consider quite extraordinary. Those crystal healers are all sitting in their fortress (in Boulder, Colorado?), and the Bigfoot glimpsers in theirs, and a seemingly uncountable number of other groups in theirs, all equally well insulated and certain about their own pet enthusiasm. They each have the "evidence" and the argument to support their own theories. They are all waiting for some other guy to come along with a battery of evidence with which they can pierce those walls. And that never happens, because we each believe that the other guy has the burden of proof.

So, what to do? First, we don't go after the other guy and his kooky ideas as much as we go after ourselves and our own kooky ideas: we each have to learn the skills of unbiased and informed analysis. (Somebody ought to write a book!) Second, we each have to accept the burden of going to the other camp to listen as well as to present the evidence we have. Indeed, extraordinary claims require extraordinary evidence, so scientists better pack every bit of data they have into supporting their shocking conclusion that Elvis really is dead. They can only hope that in listening, they will find a few other persons who will listen right back.

Idea Addiction

Does the person you are speaking with become defensive or annoyed for even suggesting he is wrong or that there is a less fantastic alternative? Physicians use a mnemonic to remember how to detect addiction, and perhaps it applies to cherished ideas as well. Whether we are talking about alcohol, mysticism, aliens in Roswell, New Mexico, or any other notion that may hold us more than we hold it, remember "CAGED":

- Cutting down and using healthier alternatives (iced tea or secular ethics) is unacceptable
- Annoyed by criticism concerning his excessive enthusiasm for it
- Guilt or serious reservation about his own use of it
- Eye-openers—needs to reference it first thing in the morning to ensure he doesn't drift too far from the influence
- Damage suffered to important issues in life due to it (declining education, jobs, relationships, and so on)[2]

Idea addiction is not always bad—Olympic athletes and other high achievers may have a single-mindedness that borders on idea addiction—but it may also be a "red flag" (a fallible indicator) that signals an above-average risk that an idea may have been taken beyond reason.

Forgetting to Track Outcomes

Beware the unintended consequences and the unmentioned flops. Doctors know that virtually every thoughtful intervention produces some intended good effects and some inevitable adverse side effects, but it's hard to reliably predict how the balance of effects will play out. The question is, was it worth it? So, too, with our efforts in public policy.

Sometimes we become so attached to the implementation of an idea that we forget that there were, in fact, some uncertainties involved and that we must evaluate the outcomes to see if the results were, in fact, those intended. Did COBRA substantially reduce the population of the uninsured? Did the Medicare prescription drug card significantly improve access to expensive drugs? Did the Patriot Act net many terrorists who we would not have caught without it? And what were the adverse side effects of any of these?

Many "solutions"—personal or legislative—are really just "potential solutions," or experiments, enacted when we had to do *something* to address a problem, despite irreducible uncertainties and unavoidable injustices. If we get the outcome's data, then we can make necessary changes to gradually improve our effectiveness.

> *The ability to come home again was essential if a people were to enrich, embellish, and enlighten themselves from far-off places. In a later age this would be called Feedback. It was crucial to the discoverer, and helps explain why going to sea, why the opening of the oceans, would mark a grand epoch for mankind. In one after another human enterprise, the act without the feedback was of little consequence. The capacity to enjoy and profit from feed-back was a prime human power.*
>
> —Daniel Boorstin, The Discoverers

Selective Scrutiny

A friend of mine is outraged that President Bush would claim that Saddam Hussein had the means and the intent to attack the United States with weapons of mass destruction but that Bush does not present clear and indisputable evidence for the claim. My friend is also certain that President Bush's emphasis on the threat from Iraq was contrived in order to create a conservative upswell that would shift Congress toward the Republicans in the 2004 elections. Yet it does not occur to my friend that he himself should have evidence for his claims, just as he demands it for Bush's claims.

An author may pore deeply into the details of the evidence that is said to support the theory of evolution. He may practice exquisite skepticism, rigorous scientific standards, and examine meticulously the thinnest slices of evidence exposing its scientific weaknesses, errors, and frauds, all to the purpose of demonstrating what seems to him an inherent scientific implausibility of the theory (bravo—all of this is commendable science so far). He then concludes that, instead of evolution, the much more plausible explanation for the origin of all life is that an omniscient, omnipotent, neuron-free consciousness quite intentionally designed and produced it all, as given in infallible scripture. Rigorous scrutiny and unrelenting skepticism to wither one theory; total abandonment of them for the preferred theory.

It is good to have high standards for proof, but such standards must be applied evenhandedly. I actually had a prospective student once say to me, "I am very interested in your class that analyzes the American health care system. I think it is self-evident that the solution to our health care problems is that we should establish a government-run nationalized health-insurance system. Can you help me find the studies that would support that conclusion?"

Sometimes we mistakenly believe that a person does not have the skills to analyze an issue thoroughly—until we present data that is counter to their preferred opinion. Then they give a thoughtful and perhaps well-researched analysis of why you are wrong. Did they put the same time, energy, and enthusiasm into scrutiny of the weaknesses of their own opinion? And into the strengths of the other opinion? Do they use their skills to scrutinize equally both sides of the issue?

> "I'd rather smoke than use those nicotine patches to try to quit because I have actually studied very carefully the risks of those nicotine patches, and there is a chance that you could get sick from using them."

Selective Scoring
Counting the hits and forgetting the misses: "Just when I was thinking about her, the phone rang, and it was her—I swear I'm psychic!" Do you suppose that the persons whom you think about frequently may be the same persons who are likely to call you? Ask yourself how many times you thought of her and she did not call or you were not thinking of her and she did call. How often then would you expect the two things to randomly occur at about the same time?

"When something bad happens, I just remember that things such as this happen for a reason—if my house hadn't burned down, I would have never moved, and if I hadn't moved, I never would have found this cute puppy. ... " A more honest assessment may admit some disadvantages, too, and there would have been other advantages if life had gone otherwise. The real scorecard starts looking more like a life of ordinary causes and effects and of chance than that some mysterious managing force is at work.

Selective Reporting (Sins of Omission)

If our research and deliberation produces a conclusion contrary to that which we preferred at the outset, then we are obligated to report our contrary findings with the same vigor that we would have had the results supported our preferred outcome. In medical research, such results are called "negative findings." It is a great temptation to ignore negative findings and to leave them unmentioned in future discussions of our more obliging evidence and preferred conclusions. This important part of rational decision making is sometimes overlooked by product researchers employed by specific business interests, students who begin a research project with a personal enthusiasm for an expected conclusion, or legislative interns "doing research" for a committed politician.

The reporting bias may not be toward an agenda, but simply toward releasing a report. When was the last time you turned on the nightly news and you just saw a still message that read "Nothing of any lasting significance happened today, and of the stuff that did, we couldn't get accurate facts. Try again tomorrow."? When was the last time your weekly professional journal arrived in your mailbox blank, but for a message that said "We received lots of manuscripts, but none of them were really up to our standards of scientific validity."? Such scenarios seem absurd to us because "that's just not the way the business world works." True. These businesses are not driven purely by a quest to provide meaningful and accurate knowledge for the maximum intellectual development of the audience; they make money by giving out whatever their huge audience demands. Some of the people in the audience may be less discriminating or have less lofty goals than you.

Any speech or report on a complex subject will necessarily fall short of being all-inclusive. Abbreviation is the norm. But when the topic is pared down strategically to misrepresent the underlying issues, it is a deceptive practice.

Encourage Independent Confirmation of Your Findings

If you are not inviting challenges, skepticism, and the double-checking of your ideas, then you are not interested in the truth. If you are offended that someone asks you for the evidence on which you formulated your conclusions, then you are not really interested in an evidence-based collaboration; instead, you are interested in the promotion of your idea. Are your findings reproducible by other impartial persons? Obtain independent confirmation early on, certainly before going public. Countless scientists have eureka pie in their faces for failing to do so.

A recent infamous example is the 1989 announcement by Drs. B. Stanley Pons and Martin Fleishman that they had achieved "cold fusion." Such a scientific milestone has long been considered the Holy Grail of thermodynamics, as it would offer thermonuclear-level energy supplies in a manageable tabletop form. The television cameras of the world gathered for their story, the contraption failed to produce, and the cameras trickled away. Independent confirmation of their findings before the big media event may have saved them some embarrassment. Indeed, if a discoverer first announces his findings to the media rather than to his own peers for professional scrutiny, we should consider this a "red flag" for pseudoscience rather than a bold announcement of reasonably well-supported science.

Balanced Presentation

A sign that a source is giving a valid assessment is the presentation of all sides of an issue—strengths and weaknesses, alternative explanations, various viewpoints, all scrutinized to the same degree. An argument that presents only one side, especially when there are obvious and reasonable considerations on the other, is not intended to elevate the level of the discussion, but to compel a particular conclusion, however inadequately informed. In addition, we can be reassured that presenting the opposite viewpoint will not reduce the credibility of our conclusion, but enhance it, if indeed the strength of the conclusion is derived by the net balance of the evidence. It was not only free speech that Voltaire advocated, but also rationalism generally when he said, "I disapprove of what you say, but I will defend unto death your right to say it."

Impressive Presentation versus Valid Content

Commercial advertising and beauty contests teach us that there is little correlation between the packaging and what lies within. This is no less true when ideas are packaged. Consider the intensity of the expressions on the nightly news, the impressiveness of meeting places, the daunting formality of scientific journals, the drama of some press conferences, the solemnity of rituals, the screamingly large print of some headlines. If they need to use such devices to get and keep your attention, you should be concerned that the producers themselves may know that the substance and analysis alone won't hold you.

At a public policy research center:

Tyisha: The evidence is clear that racial minorities in this country suffer from worse health than the majority white population. This is because of discrimination by the largely white medical establishment: hospitals and doctors' offices are not established in predominantly black areas of town; when doctors see blacks, they are treated differently and less adequately than when seeing white patients; and the physicians are not educated on the unique cultural characteristics and needs of the black community.

Jim: Those are very serious claims. What is your evidence?

Tyisha: Here are the studies.

Jim: Okay, I'll review them. Now show me the studies you have that refute these conclusions.

Tyisha: I don't have any.

Jim: They don't exist or you didn't bring them?

Tyisha: There aren't any worth discussion. At least I never found any during my research. Maybe there's something out there, but it can't amount to much if I didn't find them.

Jim: Not a single study of consideration? Can you show me the best among the inadequate studies?

Tyisha: I don't have those. Frankly, I see no point in carting around studies that are inadequately designed, biased, and contrary to the obvious conclusions of the generally available data.

Jim: To know that much about the studies, you must have gone over them pretty carefully. Can you get back to me with them?

Tyisha: Well, I didn't review them personally so much as I have heard that they are "out there," but none came up on my searches.

Jim: What databases did you use?

Tyisha: The African American Center for Injustice in Health Care, The Massachusetts Center for Racial Restitution ... They have file drawers full of studies if you go down there.

Jim: Why do you think that the studies pulled from an advocacy group would give the whole picture?

Tyisha: Because it accords with common sense and general knowledge. My grandmother hasn't seen a doctor in ten years. When an office opened near her home, she was the only African American in the waiting room. I saw them turn away African Americans at the front desk all the time.

Jim: Did they ever turn away white folks?

Tyisha: I don't remember.

Jim: Is it possible that *poor* people have worse health and worse health care and get turned away for lack of insurance and that African Americans are more likely to be poor than white Americans? Do your observations and studies account for that?

Scientific Language Does Not Mean It's Science

"Arthritis? This shark cartilage is composed of the hydrocarbons common to both these agile *Carcharodan carcharias* and your own synovium. Four out of five rheumatologists surveyed gave it two limber thumbs up."

Even well-intentioned and "serious" science can dupe us.

Leaping to Causality

> "There are very few African American students at private universities; they aren't here because they are black."

> "I won at the casino twice last year; both times I was wearing these purple socks. Now I always wear my purple socks when I gamble."

> "Marijuana causes a loss of initiative and motivation."

> "I don't do flu shots. Last year, I was fine all winter, but eventually so many people around me had the flu that I finally went in and got one. Three days later, I was sick as a dog."

Confounding Variables
In the first example above, there may seem to be a correlation between being an African American and being rare at the area private universities. But does the fact of one's race cause the absence? There may be *confounding variables*— other factors that move with race to some degree that are the more likely cause, such as economic factors (if the average African American student has less money than the average white student, then this pattern may exist entirely apart from race) or social factors (test scores, geographic distribution, cultural preferences).

Coincidence Is Not Causality
In the second example above, a person believes there is a pattern connecting two events, but there is no evidence for any connection; independent events will occasionally occur in proximity as a matter of *random coincidence*, but that does not mean they are connected to one another in a causal way.

Sequence Is Not Causality
In the final example, of flu shots, sequence is mistaken for causality. The claim that sequence proves causality—*post hoc ergo propter hoc*, "after this therefore because of this"—is a fallacy recognized since the ancients.[4]

Reversing Causality

In the example concerning marijuana,[3] there may be causality, but the evidence is not clear as to which way the causality runs—a deficiency of initiative and motivation may just as well lead to the distractions and relief of marijuana. It is also possible that the causality goes to some degree in both directions, creating a positive feedback loop. What is the fallacy of these statements?

- "I would never let my daughter be on birth control. My neighbor had this really wild daughter, so he finally got her on birth control, and a year later—bam!—she was pregnant!"
- "My two-year-old said to me recently, 'The sun is bright. That's because the snowman is melting.'"
- "There must be pressure sensors under the pavement at the traffic lights, because after I've been sitting there a long time, if I just pull up a couple feet, the light will change soon."
- "The reason our economy is so strong is that we have a high percentage of the population that holds a college degree."
- "I'm pulling for the North Carolina Tarheels in the NCAA tournament because the Republicans always win the presidency when the 'heels are in the Final Four."

Scholars used to think that the fact that human beings require about eight hours of sleep each night, usually facilitated by darkness, and the fact that their part of the globe got about eight hours of darkness out of every twenty-four were proof that the cosmos had been constructed to meet human needs. A more evidence-based interpretation is that humanity evolved within the context of Earth's rotation: some organisms found it advantageous to restore their energies under the protection of darkness.

Physicians, who have long wrestled with distinguishing correlation and causation, have relied on guidelines such as the following, which are quite instructive outside of medicine as well:

Some criteria for judging causality:[5]

1. Strength of association: the greater the relative risk or odds ratio, the more likely it is that the associations are causal. In other words, the stronger the correlation, the more likely there is actually causation.
2. Consistency: the association has been observed in a variety of populations under a variety of circumstances.
3. Specificity: the exposure causes a specific effect.

4. Temporality: the cause precedes the effect in time.

5. Dose-response relationship: the greater the exposure, the greater the magnitude of the effect.

6. Biological plausibility: there is a rational biological explanation for the association or, outside of medicine, a rational physical connection between the two.

7. Coherence: the association is in accord with the existing knowledge of the natural history and biology of the disease. Again, the connection does not require fabricating fantastic new entities or violating physical law.

8. Experimental evidence: laboratory data support the human findings.

9. Analogy: similar exposures are associated with similar effects.

Validity in the Chain of Logic

"Our kids don't do well on tests compared to the kids of other nations. ... The tests are good predictors of productive and happy citizens ... so we need to gear our high school courses toward the material seen on the tests ... so we need to revise school curricula and reduce workload for individual teachers ... so we need to give the schools more money. ... " Identify each link that is involved, consciously or subconsciously, in the chain of logic that is being used. Is each link sound?

False Endpoints

Keep your eye on the goal.

Physicians are frequently assured by the pharmaceutical salesperson that her newer, more expensive antibiotic is clearly superior to the older, cheaper one because "the colony count of bacteria in the culture dish diminished faster with the new drug." Hmm ... how sure are you that such a phenomenon in a culture dish reliably indicates that your patient will actually recover more quickly? Could it be that the environment in the dish is not quite the same as the environment in the throat? When she is asked for the actual study that shows a sufficient number of humans diagnosed with the infection for which these drugs are intended, with a random half taking the old drug and the others the new one, then compares genuine clinical outcomes, she may become less confident in her assertions; thus, she gives you a false endpoint—a colony count of bacteria in a culture dish—and wants you to believe that it means your patients will feel better sooner. Quite often, they don't.

False endpoints are also common in public-policy discussions. In the ongoing debate concerning race-based admissions policies for universities, the law of the land had been established in 1978 in the U.S. Supreme Court decision in the University of California versus Bakke. In that decision, Justice Powell accepted the university's rationale that a racially or ethnically diverse student body enhances the educational experience of all students. The evidence that this is so remains contested, but, for now, it is the ultimate rationale for affirmative action. However, many other rationales for affirmative action were found unconstitutional: Do the students admitted with the assistance of affirmative action policies get high grades in school? Do they earn high incomes after school? Do they become leaders in their field? Are they more likely to have children who get into college on merit alone? Do the racial majority students demonstrate less ignorance or racism toward the minority after the shared educational experience? All of these are interesting questions, but they neither support nor oppose the notion that affirmative action policies enhance the educational experience for all students. They are false endpoints relative to the ultimate criterion.

Keep in mind the ultimate purpose or the original claim and hold people accountable that their evidence actually supports it and not some other presumed and possibly false endpoint.

Valid Conclusion, Invalid Application

This is a special case of the preceding notion and deserves special attention.

Suppose New Jersey policy makers were assured that studies had shown no correlation between speed limit increases and highway fatalities. The New Jersey turnpike increases its speed limit then has a terribly different experience. A second look at the study shows it was done in Wyoming. (I made that up, Jersey commuters.)

For many reasons, we may regard some evidence too highly, and we may apply it where it does not apply. For example, our vaunted randomized, controlled trials that seem to provide us with concrete evidence about cause and effect for an isolated variable may well do so, but only under the specific conditions in which that trial was performed. Those exact conditions may not often be present in the real world, and it remains unclear if the differences are enough to render the previous lesson irrelevant. A medical study on the effect of a medication on high blood pressure may have been done with men only ... and who were white only ... and who did not have heart disease ... and who were on no other medications. Is the lesson learned relevant to an older African American woman with heart disease?

Scientific studies aside, the valid conclusions of life experience can also be misapplied. Years of experience in business in Washington, D.C., may convince you that a dark suit and conservative silk tie is the way to be trusted, but when you wear it in Durango, Colorado, your audience shuns you. As another example: several months of dating convinced you that a man respects financial prudence, but his birthday, you discover, is a special case.

In the real world, there are many variables affecting a situation—have we considered them all? Do we know the degree of influence of each? Can we predict their collective effect? Why is each of these statistics meaningless?

Misunderstanding Statistics

Understanding statistics is a critical skill in weighing the validity of scientific studies, which is why the topic is mentioned here. However, chapter eight is devoted exclusively to the top ten lessons that everyone without a degree in statistics still needs to know.

Statistics of Small Numbers

- "It's going to be a landslide for the democratic candidate—at lunch today, all five people at the table said they voted for him."
- "Muslims are such irrational and angry people—on TV they're always screaming."

Variables May Be Clumped

- "They say one out of five Americans is Latino, but I know hundreds of people, and none of them are Latino."

Independent Events

- "I flipped this coin five times, and it was heads every time! It *has* to be tails this next time."

Confounding Factors, Causality

- "In a big study, only 2 percent of the twenty-one-year-olds grew up listening to Mozart, but among college graduates, 20 percent did—Mozart must make you smarter."

Economic and Education Confounders
(See also Suppressed Evidence and Half-Truths on page 174.)

- "This table shows that the counties with more black people have higher rates of infant mortality. That means we are giving worse care to people because they are black ... "

Modality
- "Georgia may get more rain per year, but it comes in brief deluges, whereas in Washington it rains more steadily, with less intensity."
- "I'll do my weeklong hike in Washington State rather than in Georgia, because weather studies show it actually rains more in Georgia."

The Bell Curve of Normal Distribution
(Averages may come from broad bell curves; two broad bell curves may be slightly offset but still overlap so much that the differences within each group are far greater than the differences between the two groups.)
- "So all thirty-two of us sat in class trying to predict if the coin would be heads or tails. ... I flipped the coin five times ... this one guy was right every time! He's psychic!"
- "You have to go to Nebraska to buy a car—the average price of a car in Nebraska is $500 less than in this state."

The science of statistics is amazingly powerful. Well-done statistical studies can be enormously helpful in understanding complex problems and their underlying factors. These studies are so elaborate and flexible that they can almost seem as organisms themselves. But, similar to organisms, they can be misused. They can be dissected into parts, cut up into little statistical quotes that individually represent the original whole no better than a single bloody piece of tissue tells the lay person much about the organism from which it was excised. The beauty and usefulness of the original whole can be shot to pieces, the most impressive fragments put on display in ads and opinion pieces to serve some end not necessarily consistent with what the study in its entirety would support.

Individuals who work with physical organisms, including physicians, know to be extremely cautious with such fragments. They have an honored adage for their work: "If it's wet, and it's not yours, don't touch it." There may be wisdom in this that is applicable to the handling of statistical fragments: it's wise to be disgusted; keep your distance; use the mental equivalent of gloves and tweezers; and, most importantly, ask if you can't instead see the whole study from which it originally came, still intact and functioning as its creator intended. But if you ignore *all* statistical evidence just because it is easily manipulated, you also lose out. The key is to be knowledgeable in statistics so that you can accurately ascertain whether particular statistics are reliable or not, on a case-by-case basis.

Bones of Aliens

This is the claim of irrefutable proof that is never actually in hand. We've all heard the breathless stories about the person who found in some remote place the twisted wreckage of some unusual craft, made, no doubt, of a mysterious and unidentifiable metal. Within it was found the remains of some type of man who was, well, not quite a man. These great stories are always unverifiable because the actual hard evidence of the story—the wreckage and the bones of the alien—are always just out of reach of authoritative scrutiny. Nonetheless, the story always comes from persons who are very convinced of what they have seen: "But I saw it with my own two eyes!" Or, "I didn't get a good look at it myself, but I know this guy whose cousin was there!" Sometimes the person will indeed produce some fragment to "prove" his story, but those fragments always turn out to be much more mundane items: an old car part, a weather balloon, a bone of a cow, a blurry photo of an airplane. It is human nature to be so enthusiastic about the Big Discovery that our imagination is triggered and we overinterpret a small item.

No scientist should expect to be believed simply because she states that she has seen studies. Similar to the bones of aliens, studies that "prove" an especially shocking discovery or unorthodox claim probably don't exist, or are not what she thinks they are, or were overinterpreted far beyond what can be reasonably and honestly understood from them. Occasionally, however, some of those surprising claims actually turn out to be valid discoveries, and we have to rewrite the science books. Be among the first to know the new findings—settle for nothing less than the original research studies and know how to evaluate their validity.

In summary, the concepts and methodology used by scientists to think clearly and logically in order to understand phenomenon can be employed by anyone, whether they consider themselves "science types" or not. No one needs to put on a white coat or protective eye goggles to begin to think methodically; to consider all explanations; to research, evaluate, and report in an unbiased fashion; and to admit degrees of uncertainty so that an even better answer can be later formulated—all regardless of one's preferred beliefs.

While science, with both its successes and failures, provides some lessons on how we can all think more rationally, religion does so as well. We turn now to that timeless aspect of the human experience.

Chapter Four
Religion, Tradition, and Moral Codes

The God Reflex

The God Reflex is humanity's tendency to attribute anything not understood to God or gods. It is another example of the errors that result from our reluctance to simply say, "I don't know." The things not understood may well be powerful (such as natural laws or germs), but they are not necessarily cognizant beings. The uncertainties may address daunting questions for which the answer is difficult (such as, "What is the ultimate origin of our moral code?"; "Can we ever have justice for everyone?"; or "What caused the big bang?"), so we may choose to fill in the gap with an omniscient being. Not only is there a human tendency to employ gods as placeholders where we have no other explanation, there is also a human tendency to anthropomorphize those forces, to give them the shape of man. Since we tend to judge ourselves as the highest form of life, we can find no better model for the gods we invent. So, from natural, unconscious, physical laws, we create gods, and then we endow them with bodies and faces and, more significantly, with consciousness and concerns similar to our own: people created God, and we created Him in our image.

Why does man live? Chorus: God.
No, it is his heart that beats. (Empedocles, 470 B.C.E.[1])

Why does the heart beat? God.
No, it's electrical impulses from the sinoauricular node. (Kollicker and Mueller, 1855)

Why does this node produce an impulse? God.
No, it is the electric potential across the membrane of a specialized cell.

Why does the membrane do this? God.
    No, evolution. (Charles Darwin, 1858)

Why should there be evolution? God has methods too.
    No, slight variations of organisms and the struggle to survive.

Why would there be variations? God.
    No, DNA mutates. (James Watson and Francis Crick, 1953)

Why does DNA mutate? God.
    No, natural background radiation. (Salvador E. Luria, 1943)

Why is there radiation? God.
    No, the laws of physics. (Isaac Newton, 1687; Marie Curie, 1896)

Why do these laws exist? God.
    No, all phenomena are mechanical products of the big bang.
(George Gamow, 1968)[2]

Why would there be matter or energy from which a big bang could
occur?
    "I don't know yet, but this pattern suggests the eventual
answer won't be 'God.'"

The progress, however painfully slow, in answering those questions was
possible only because some persons did not join the chorus; they resisted the
God Reflex, instead answered, "I don't know," and pursued a rational and
natural explanation. To give up, simply draw in a man's face as the cause of
the big bang, and declare the problem solved is anthropomorphization. Even
if we did—for the sake of discussion—accept that answer, we would pursue
it as we had every other answer; we would have to ask "Where did God come
from?" and the answer is typically something to the effect of "God just
always was." Well, if it is logically acceptable to say, "God just always was,"
then it is fair and logical to say, for now, that until we have better data, perhaps "matter and energy just always were." Furthermore, the answer requires
no violation of physical law, no creation of a new entity to jerry-rig the
explanation (à la Occam's Razor)—it is the simpler answer that explains the
observed phenomena and keeps us looking for a more specific explanation.
    Many people standing on a mountaintop at sunset or peering down at

a newborn baby in their arms will have a sensation of indescribable awe, gratitude, connection, humility, and inspiration. It is a real and valid sensation that may occur in many other profound circumstances as well. To ascribe it automatically to an omniscient being is anthropomorphization. It may well be beautiful, awesome, and mysterious, but that may well be because nature is very capable of beautiful, awesome, mysterious acts. You may be created by it and inseparably connected to it, but in more of an unintentional chemical way than in an intentional personal way. We have rational evidence for the mechanics of nature, but we have no rational evidence for putting a man's face or a woman's face on it and saying that it was done with conscious intent—that is the God Reflex.

It takes nothing away from the spectacular and moving nature of these experiences that nature does this on its own; indeed, the realization of this often elevates a person's appreciation of the material universe. To be alive and to experience the stunning and exhilarating grandeur of existence in the cosmos is genuine; to anthropomorphize it is not.

We continue to respond with the God Reflex in many circumstances today. Persons who have sufficient education on the science of each of the principles involved at each of the evolutionary steps mentioned on page 121 will usually accept the science and perhaps even shake their heads in wonder that there still remain some who do not. And yet, the same people, when thinking about the new issues, may nevertheless follow the same reflex: scientific answers from the distant past are fact; scientific answers from the recent past are theory; scientific answers in the future are heresy. The God Reflex seems hardwired in us.

There are many examples of the God Reflex. We see the God Reflex on the edge of ancient maps, where the known world fades into uncertain territory. Here the cartographers have filled the blank periphery with illustrations of the beasts of imagination and of God himself, as if they patrolled the borders of the (known) world. Through exploration, our knowledge of the world increased and the maps expanded, but one thing was especially slow to change: the unseen territory that lay beyond our clear knowledge was populated by us with gods, demons, and mystical creatures. Although they were always pushed back out of sight by advancing human knowledge, we seldom failed to fill the space just beyond with our gods of the time. When exploration of the globe was complete, the cartographers turned their attention to the space beyond the globe, and, undeterred, we mentally penciled in our God just beyond that space too. Even today, when we want to look to our God, we look to the sky, still quite certain he is there, just beyond our observable universe. Many recognize the unlikelihood that this God would exist in bodily form ("How could he survive in the vacuum of space?"; "Why would he have lungs before there was any air to breathe?"; "Or opposable thumbs with nothing to grasp?"), yet we still grant to our God the ultimate anthropomorphizing qualities: consciousness, intent, and concern with human affairs, especially one's own individual affairs.

To this day, insurance contracts refer to natural catastrophes as "acts of God."

The ancient Greeks attributed any surprising or abnormal behaviors to the *moria*, or *ate*, which had been put into the person by a god to suddenly fill him or her with the courage, confusion, or passion to perform particular acts. These acts we now attribute to adrenaline, ethanol, or sexual hormones. The ancients knew nothing of these agents, but they knew well their sensation and their effect and so were left to answer that the immediate cause must have been "God."

This sentiment remains powerfully with us today. The word "enthusiasm" comes to us from the Greek *entheos*, meaning "having a god within," which we still see in our expression "Something came over me." Consider, also, the origin of "to be inspired," which is literally, "to take in the spirit." The word "pain" comes from the Latin word *poena*, which means punishment or penalty.[3]

Some persons have more than nature and nurture in mind when they say a smart child is "gifted." In *The Greeks and the Irrational*, E. R. Dodds reminded us that this is the origin of our diagnoses of "stroke," "seizure," "attack," and "touched in the head." We still refer to alcoholic beverages as "spirits," and a person who behaves in an extreme fashion is considered to be "possessed." Historically, many cultures did not distinguish between their religious men and their doctors, because without the germ theory, what seemed to have entered and thus required eradication were conscious and intentional demons. Here again we see the human tendency to anthropomorphize.[4]

It has been a disappointment—and sometimes a great irritation—to many through the course of scientific discovery that not only can *causes* of phenomenon be explained by worldly agents, but that those agents are not conscious and intentional; they are blind, *mechanical* systems of nature (electricity or genetics, perhaps) or lowly organisms (a bacterium, perhaps). People strenuously resisted Nicolaus Copernicus's revelation of a sun-centered solar system, Charles Darwin's theory of the evolution of man, Sir Charles Lyell's theory of the actual age and development of geologic features, and the gradual scientific dismissal of one cherished myth after another.

Such discoveries are exhilarating to some but are directly threatening to others. Today, many people refuse to consider natural, real-world explanations for other phenomena, such as the notion that their cherished religion has at its ultimate origin not a divine mover, but borrowed myths and legends of many other religions, useful and instructive fabrications, beautiful and comforting hopes, all ultimately made by normal men and women.

Certainly, there is much about the universe and the human experience that cannot yet be fully explained—many questions remain about the big bang, consciousness, and the future of our planet. There are "places" that our explorers have not yet gone: the limits of deep space, subatomic realms, and through time. Thus, even today, when scientific explanations are exhausted, the last questions are often still answered with "God." Nonetheless, if we wish to follow the truth wherever it may lead us, we must not give in to comforting and mystical answers; we must resist the God Reflex, admit that we do not know, learn from history, understand why it is in human nature that we create fictional answers, and, finally, set ourselves to learning the truth.

If we presume that most apparent irrationalities actually have their own underlying but perfectly sober reasons, then what are the reasons for the God Reflex? First, as discussed above, we dislike saying that we don't know. As nature abhors a vacuum, so, too, does human nature abhor that long pause when someone asks an unanswerable question. Many of us can't tolerate not having an answer, so, right or wrong, "God" serves as a temporary placeholder and gets us off the hook. However, our sense of satisfaction at answering the question is unearned.

Second, our memory of our track record with this answer isn't very good—we can still claim that the immediate cause of some things is God with a certainty undiminished by our long history of being wrong with that answer.

Third, we are genuinely fearful—we sense that this pattern of factual discovery will, in fact, go on forever, and that as it does, the pattern of ever increasing diversity, challenges, and incomprehensibility of the universe will also continue and the size of humanity (and certainly of self) in relation to it all will conversely and dishearteningly shrink. These diversities and challenges the old cartographers rendered as menacing dragons. They anthropomorphized the limitlessness and incomprehensibility of the universe with the face of an omnipotent God.

Fourth, we may create these creatures as an alternative answer to what we worrisomely glimpse as a likely but unpleasant answer: after death, there may be only oblivion; before the big bang, perhaps there was nothing but eternal faceless energy; your life does not have a predetermined purpose, but it is up to you to give it purpose. Confronted with such evidence, many of us prefer the idea of a comforting god.

Fifth, we confuse rejection of traditional answers with rejection of the love or respect we have for people and institutions in our lives. People have a responsibility to the truth; they can retain the same affection for the people in their lives while choosing to disagree with them. If a relationship is challenged by a divergence in opinion, let the relationship rise to the challenge by having open and civil discussions of the issue. Reassure the other side that you do not have a lower opinion of them if they do not agree with you. It underestimates the relationship to assume it is contingent on comforting agreement.

Last, the God Reflex reinforces itself. Once having used "God" as our placeholder answer, we quickly forget that, in fact, we really don't know, and come to believe that "God" is actually the full and final answer. We have thereby inserted an answer that may be so intimidating or culturally revered that we and others will be very slow to investigate further. Subsequent inquiry and research are less enthusiastic, if not actively suppressed.

That people would have a reflex that is not entirely healthy or that can lead us into trouble is not at all unusual. Anger, suspicion, and jealousy are just a few of the other natural human responses that serve us in some ways and yet can also be inappropriate and cause us a great deal of trouble. It is a matter of individual and social progress that each of us learns to control and manage those reflexes when they arise, expressing them when appropriate, and repressing them with more cerebral considerations when they are not.

Even today, too few persons acknowledge and manage their God Reflex as does our speaker in the final line in the dialogue above, despite the pattern of science and history; thus, the God Reflex earns a special and ignominious place in human psychology and in the history of reason.

The advance of reason and the human progress it brings depends, in part, on people acknowledging and managing the God Reflex. Why? Because critical analysis and the skills of reason aim not just at answers, but at the *correct* answers, no matter how sensitive the questions or how cherished the preferred answer. As we saw in the discussion of objective versus subjective questions, the question of whether an omniscient being exists or not is an objective one—regardless of our preferences, there is a correct answer, so we must use valid evidence and reason to learn what it is.

Confusing the God Reflex with Religion

It is a fallacy to believe that by criticizing irrationality, one is dismissing religion. The irrational aspects of a religion—superstition, unconditional faith, or the belief in any violation of physical law as examples—are only parts of the much more complex institution. There are many other aspects of religion that are profoundly important to civilization and the human experience—education, community, morality, beauty, awe, inspiration, hope, gratitude, forgiveness, charity, and consolation, for example—that can, and should, be maintained. They could, perhaps, be cultivated in rational religions—religions formed or reformed to advance all those noble aspects of the human experience without ever resorting to superstition or violations of physical law.

Religions and the people who follow them have made noble changes to reduce their irrationality—we no longer burn "witches," for example, despite the command "You shall not allow a witch to live." (Exodus 22:17). After having executed 100,000 persons through history for this nonexistent crime, there were persons who thought that abandoning such an apparently important belief and practice of the religion would be not *reforming* the religion but *demolishing* it, and calling it the same religion would be disingenuous if not absurd.

The giving up of witchcraft is in effect the giving up of the Bible.
—John Wesley, founder of Methodism, circa 1760[5]

And yet, there were others who felt that by instead admitting error, stepping around the awkward phrases in their scripture, and doing what actually seemed rational and consistent with a more enlightened moral code, the ultimate purpose of their religion was not just preserved but, in fact, *better* served.

A second example is seen famously in the eighth century when Leo III, a pious Byzantine emperor, took Deuteronomy 4:16 literally and issued an edict in 726 that his soldiers were to remove or obliterate all icons of Mary and Jesus from churches. Statues, murals, frescoes, and mosaics were removed from the public eye. In city after city, common people, low-level clerics, and soldiers refused to submit to the notion—the Bible's or Leo's—that these images did anyone any harm. The rebellion rose into the upper levels of the Church until Irene, empress of the East, and 350 bishops reversed Leo's decision and ordered the images restored to churches throughout Europe.[6/7] Deuteronomy was preserved, but human judgment prevailed.

There are many other examples of religion gradually setting aside its manifestly irrational tenets and its claims that now seem to be physically or morally absurd. It is a process that needs to continue, wherever that may lead us. It is possible for communities to continue their great tradition of serving their noblest aims without resorting to fallacy and fantasy.

Religion is not under special scrutiny here; irrationality in all forms, including the God Reflex, is. Scrutiny is applied everywhere, and in this book, we focus particularly where irrationality is detected, to whatever institution or person exercises it. If irrationality is found in religion, science, education, government, business, or the family, it will be scrutinized. If it is found in our moral codes or in the stories of our heroes real and imagined, it will be scrutinized. Irrationality in any form diminishes the integrity and value of all these institutions.

The Age of an Idea Does Not Increase Its Validity

> *… a long habit of not thinking a thing wrong gives it a superficial appearance*
> *of being right, and raises at first a formidable outcry in defense of custom.*
> *—Thomas Paine, an American revolutionary and writer,*
> *opening line of the introduction to "Common Sense"*

We keep old ideas around because they are useful to us, not necessarily because they are true or just. Perhaps it was a lucrative idea (such as snake oil and other inadequately tested products for which there was no more effective alternative), a self-protective one (economic discrimination), a comforting one (communication with the afterlife), or a validating one (Are you saying that my dear parents and all my ancestors before them were *wrong*?!), but it wasn't necessarily true. Traditional and cultural ideas may be defended as true and just beyond all reason, however, for the sake of these other benefits. Henry David Thoreau often marveled, for example, at the cultural rules and expectations of his day—in this case, as concerns the timeless but exhausting and frustrating race for material accumulation, as he described in his essay "Economy."

> *When we consider … what are the true necessaries and means of life, it*
> *appears as if men had deliberately chosen the common mode of living because*
> *they preferred it to any other. Yet they honestly think there is no choice left.*
> *But alert and healthy natures remember that the sun rose clear. It is never too*
> *late to give up our prejudices. No way of thinking or doing, however ancient,*
> *can be trusted without proof.*
> *—Henry David Thoreau, "Economy"*

Sometimes it is not the tradition itself that is important (and, in fact, it may seem on the surface to be so archaic and absurd as to seem ridiculous—consider the Easter bunny), but it is the *underlying ideas* that matter. While the historical accuracy has long since been disproved, the traditions remain as the vehicles for reminding us of ideas that are still relevant and profoundly important.

> *It is an unscrupulous intellect that does not pay to antiquity its due reverence.*
>
> —Erasmus, *a cleric, humanist, and*
> *philosopher,* Works of Hilary, *1523*

Can you change the outward vehicle and maintain the critical underlying ideas? If so, why bother? Why not just enjoy a useful superstition?

We may bother for several reasons. We may bother when the tradition itself eclipses the actual meaning and intent of the tradition. You have to decide for yourself what meaning and intent you celebrate (apart from what everyone else expects you to celebrate), then decide if your tradition and ritual facilitates that purpose or distracts from it. We may bother when people force themselves to hold a traditional belief to be literally true when it is patently absurd and thus threatens to teach irrational and mystical thinking as a valid approach to real-world issues. We may bother if a tradition that relies on ultimate justice in the next life slows us down from demanding and constructing a better system of justice in this one. We may bother when, in the name of cherished traditions, people inflict injustice, discrimination, and unequal treatment under the law and restrictions of freedom on persons who don't share the tradition. We may bother when the tradition is explicitly hostile to critical elements of civilization such as education, health, social toleration, and human rights.

A Book's Age Does Not Increase Its Validity

Determining today what transpired in the past is very difficult. There are two general types of evidence: material evidence and human testimony. Material evidence is largely a matter of scientific method: geology, archaeology, paleontology, carbon-dating, DNA testing, fingerprinting, forensics, and so on. Human testimony is the account that persons give us directly or in some medium such as a written record.

Human testimony is among the weakest forms of evidence because there are so many ways in which humans can knowingly or unknowingly alter the story. The mind can play tricks so that what people perceive is quite different from what actually occurred in front of them. They may elaborate on some points, omit others, dramatize, or understate. They may insert "clarifications" that only muddle. They may confuse and combine the events in question with wholly unrelated events in previous or subsequent experience. They may make errors, introduce well-meant but invalid additions, or commit outright fraud. These alterations are accelerated and magnified by emotion, time, number of transfers of information from one entity to the next, number of reproductions of the data, lack of original data, lack of communication between persons with differing accounts, and political or personal incentives, such as status, expertise, or ego. Thus, real and historical events may be launched into the realm of legend with persons, events, discussions, and purposes so completely different from the originals as to bear no resemblance to actuality; the historic persons involved in the actual events might well be appalled and amazed at how the future would know them.

The divergence of the myth from the historical fact may begin at the very moment of the actual event itself, depending on how methodically and thoroughly the event was recorded. An event may be well documented by a variety of disinterested independent sources or perhaps by a fewer number of sources highly regarded for objectivity and nonprejudicial treatment of many previously well-substantiated events—if so, a sense of historical accuracy may be warranted. If not, then the actual events quickly become shrouded by time. There may be ways to penetrate that shroud, as with fallible but compelling scientific methods or by deducing from observable consequences, but these are less reliable than properly collected firsthand observations of impartial witnesses. In our own time, we often find that a well-traveled story is upon us, and determining if it is truth or legend may be impossible. There are a few basic tools for measuring relative plausibility, however:

1. Does the story require supernatural events, entities, miracles, or any other suspension of known physical law?

2. Are the "witnesses" firsthand witnesses? Is the information secondhand or much more distant than that?

3. Is the witness impartial? Does he or she have an emotional or material attachment to your final decision?

4. Are there equally valid or more valid witnesses who have a different understanding of the events? Are they heard and considered without prejudice?

5. Does the story provide bits of information that are mundane as well as the bits that are dramatic? Ambiguous as well as incisive? We should be skeptical that we are hearing unfiltered testimony when all the information seems to point to a single conclusion, especially when that conclusion is quite extraordinary.

Scripture around the world, while much revered, does not hold up well to this type of scrutiny. Nor, for that matter, do many respected secular works of history. But the credibility gap is filled in for the scriptures with blind faith: "The actual history of the development of the book may have holes, it may have every characteristic of a purely human effort, but I am nonetheless certain that God saw to it that the book before you is the one true reflection of His message." Again, the God Reflex.

Why Do Traditional Ideas Sometimes Seem to Last Beyond Their Practical Validity?

Some of today's sacred traditions may be astonishing, humorous, or harmful preservations of what were once simple and pragmatic behaviors. For example, perhaps the prohibition of eating a particular meat in some cultures came after a gradual awareness of an increased rate of gastrointestinal illness associated with that food and of a decreased rate of the same with abstinence from it. With no knowledge of the microbial world, the culture might have decided it was God's obvious disfavor that wrought such discomfort and thus forswore its families from eating that meat. It worked—fewer people became sick. But after Louis Pasteur and refrigeration, it may be debatable whether the tradition retains its original value. Yet to change now means we are risking the wrath of God. If it's a choice between eternity in paradise and a ham sandwich, this tradition won't be going away soon.

In fairness, let us consider that this tradition—and many other apparently irrational rituals—may still be observed today not for its original purpose, but now as a symbol, if an arbitrary one, of greater ideals. Abstaining from pork symbolizes one's willingness to give up something of value or pleasure for the sake of the many ideals one's god represents. That's rational, so long as the sacrifices are still worth the benefits to you and you realize that the ritual itself is no longer the critical factor.

Another example: perhaps the sanctity that many religions ascribe to chastity and monogamy grew, in part, from the observations of the adverse consequences of pregnancy in underdeveloped girls, single mothers, and polygamous families and the transmission of sexually transmitted disease, as well as many other potential hazards. Perhaps technologic innovations such as contraception, condoms, safe abortion, and antibiotics have made some traditional restrictions less critical. While there may remain many good reasons to encourage traditional sexual mores, the point is that the traditions themselves are not the ultimate purpose, nor is their origin necessarily divine. The underlying pragmatic benefits are the goal, and the traditions that promote them are the result of astute human observation and reason. The active link between the behavior and the adverse consequence no longer needs to be filled by the typical placeholder, "God," but has been shown in many cases to be physiology, economics, and microbes. Perhaps the people who choose to break the rules need only tangle with those and a few other earthly realities, not a voyeuristic and vengeful omnipotence. If those concerns can be otherwise adequately addressed, then continuation of the restrictive, neurosis-inducing, and sometimes-terrifying traditions is reasonably debatable.

If the reasons for a tradition don't seem valid to us anymore or if the cost of them is too high for the benefit, we can reasonably consider change. We must change them with care, however, because they may be a product of generations of experience, with benefits we do not fully comprehend. Will we substitute in their place something that offers the same or more benefit with less or none of the cost? On the other hand, some traditions may be simply obsolete or a product of generations of ignorance, fear, or gross irrationality and are acting now to perpetuate the same for yet another generation. We must each think for ourselves and decide for ourselves. Carefully.

Claims of Morality

"But that's simply immoral!" That's a serious charge, often intended as a showstopper. If an action is "simply immoral," what more discussion could there be?

Plenty. A claim about the morality of an issue is not the end of the discussion; it is the beginning, because what is immoral to one person is not necessarily immoral to another. The "immoral" act may even be admirable to some, as with taxation for social services, euthanasia, a preemptive military attack, or using cloned stem cells to cure a person's catastrophic illness. When a person declares an action immoral, she must immediately explain *why* she considers it immoral. And the answer usually comes from one of two types of moral codes: a religious one, which usually claims it is a timeless, unchangeable code received from the universal omniscience, or a humanistic one, which claims it is a code derived from one's best judgment of how to balance conflicting interests according to real-world considerations, such as life, security, and property, and which is continuously revised according to new knowledge and insights.

Morality and religion are distinct entities that may or may not occur together. People can be very religious and very immoral (sadly, there are innumerable examples in history of slaughter and mayhem by devoutly religious persons), conversely they can be very irreligious and very moral

(Friedrich Nietzsche was a very conscientious moralist and atheist). Nonetheless, many religious persons make no distinction between their religion and morality, so they make the mistake that criticism of their religion or of their religious moral code is an attack on morality itself. A rational person must keep in mind the distinction between religion and morality.

There is nearly universal agreement that morality has a critical role in law and public policy; there is nearly universal disagreement on whose morality that should be. There remains vigorous debate on the role of religion generally, apart from morality, in government. A rational person must answer to all three of these discussions.

The Origin and Evolution of Moral Codes

Despite assertions by some groups and traditions that their moral code is absolute for all peoples and all ages, there is no valid evidence for any other explanation but that all moral codes are purely human constructions, created by ordinary mortals, with no supernatural input whatsoever. The founders of these moral codes may have preferred, for many reasons, that people believed that their moral code was delivered from omniscience, but their preference and insistence does not make it so. Nonetheless, we are assured that Hammurabi received his great code from Shamash; Moses's from Yahweh; Muhammad's from Gabriel; Manu's from Svayambhuva; Joseph Smith's from an angel. ... The followers themselves may prefer to believe this is so—again, that does not make it so.

Why would founders and followers prefer a divine origin? Perhaps because this cows people into rapid adoption of and obedience to the code. Or because it seems to definitively settle any questions raised by dissenters, ambiguous ethical dilemmas, changing circumstances, or new discoveries. Or because we feel special having a direct connection with and knowledge of the highest power in the universe. But this power, certainty, and satisfaction is unearned, and it is an avoidance of the personal responsibility to grapple with difficult and conflicting ideas.

Moral absolutism may actually become an obstruction to implementing better solutions. This is the "Frankenstein Effect," to which we referred earlier—when what was built by people to serve people instead becomes a powerful entity that demands *we* serve *it*. Social problems (think again of the examples above of unintended pregnancy, illegitimate children, and sexually transmitted diseases) are addressed by identifying behaviors that increase or decrease the social problems (sex anytime I and any willing partner desire, or sex only within marriage, and so on) at a relatively small cost (frustration of human nature, guilt, or fear of divine retribution), and those behaviors become part of a moral code for behavior.

As time, society, and technology advances, new solutions are discovered for those problems (birth control, condoms, abortion, antibiotics). This feels threatening to the institutions (some churches, perhaps) that have for so long made the old solution—the traditional moral code—such a part of their identity. Thus, the churches oppose the new solutions even though one would expect them to embrace them as helpful in fighting the original problems: "Giving condoms to teenagers makes it easier for them to have sex outside of marriage without suffering the consequences of disease and undesired pregnancy, so they may choose to have sex when otherwise they wouldn't have, so giving condoms to teenagers is bad." This is the Frankenstein Effect again: suddenly, these helpful new innovations are themselves denounced as immoral, as if the whole point were to preserve the church traditions rather than to preserve health and security. The moral code and the institution— originally built by people to solve specific problems—become forces that demand people serve the codes and the institutions instead, regardless of the original problems.

In a free society, people may certainly navigate by whatever moral code they wish, within the confines of the law, but it is the process of making that ultimate law for a society with a plurality of religions and moral codes that becomes tricky. Basing law on any code that considers itself absolute and unchangeable risks serious injustice to those of other codes and to posterity.

Not all religious traditions are dogmatic, and *most*[8] are less absolute than their faithful may realize:

> *Thus I have explained to you knowledge still more confidential. Deliberate on this fully, and then do what you wish to do.*
>
> —Bhagavad Gita, 18:63

Consider whether you believe that it is immoral for a woman to walk in public in slacks and a blouse rather than being covered head to toe in a burka. Is it immoral for a person to have sex with another person of the same gender? Is it immoral for a person to steal food? Is it immoral to steal food under all and any circumstances? Why?

Free societies that allow for a plurality of religions and for separation of church and state must confine their definitions of morality to the concerns of the real world; to define morality according to whether of not an action offends one's god is insufficient, since the next citizen can just as certainly believe that the same behavior pleases their god or that the denial of it offends their god. But in the real world, we can collect *material* evidence to measure *material* harm and benefits associated with alternatives. Policy makers must ask: "What real-world risk or harm is there with this action? What real-world risk or harm is there with the alternative?" We need valid evidence of the *material* danger to people or property or of genuine substantial risk of such harm before we pass laws with serious consequences, such as restricting the liberty of citizens. When we make real-world consequences rather than supernatural approval the ultimate goal of our moral code, it is called a *humanistic* moral code.

In humanism, behaviors may be *ethically* good (helping a little old lady cross a snowy street) or bad (ignoring her and leaving her to struggle alone), but they become *moral* issues only by matters of magnitude (shoving her into oncoming traffic). Humanistic morality is not defined as a central concern of a deity, but by its relative seriousness—the net weight of its consequences, good and bad, in the real world. If the consequences of the behavior are real, observable, and overwhelmingly damaging to the goals of the society, then the behavior may be considered to reach the level of immorality (child abandonment). If the consequences are real, healthful, and overwhelmingly constructive toward those goals, then the behavior may be considered to reach the level of a moral necessity (sanctity of contract). This type of morality is not usually absolute, but depends on particular circumstances, new data, or practical considerations. Thus, secular morality may change with changes in society, knowledge, and circumstances. The morality of an issue is considered open to debate at all times; it is the debate itself that reveals what real-world data is needed and determines morality of the behavior.

But such complexities and relativism are unsettling to many people. They wish for the simplicity and immutability of absolute universal law. For that reason, they may seek refuge by saying, "I don't know why I'm against it. It's wrong. It just is. It's immoral, that's all." Plato is one of many examples of historical persons who, out of frustration for lack of knowledge, real-world data, and interminable disagreement with his preferred solutions, leapt over practical considerations and declared actions inherently and indisputably moral or immoral. We see such cognitive leaps, or "Platonic breaks," frequently in contemporary debate. It is a form of the God Reflex: when you run out of ideas, declare it settled by God. For the rational person, that answer is not good enough.

Again, we must not misconstrue criticism of the irrational aspects of religion as a criticism of the good, healthy, rational aspects of religion. We are focusing on the irrational aspects of some religiously oriented moral codes. However, for many persons, religion is the dominant source of information and stimulation for their ethical and moral development. Indeed, these persons often believe that to lose religion is to lose morality (see "Fear of the Mythical Void" on page 171). It may be best for society if these people retain even the irrational aspects of their religion, if they are not capable of developing a healthy moral code without them. Perhaps the moral codes offered by religion can serve as a minimum standard below which none should fall.

But other persons may wish to attempt to improve on religious moral codes and aim even higher. Some believe that doing so is a fundamental personal responsibility of adulthood.

Think for Yourself: Describe the Moral Code You Live by

A very important exercise that I strongly encourage all my students and all readers of this book to do is to sit down and write out what they consider to be their own personal moral code. Point by point. "Thou shall not kill" may be one, for example. But are their any exceptions to that? Is it possible for there not to be? What are all the exceptions? Why? As you write down as many specific "rules" of your moral code as you can, look for patterns: What are your guiding principles? Do you offer more flexibility for persons of your own circumstance and less for people of entirely different circumstances? Do it again in three years or fifteen years—did it change? Why? Is that okay? How do you feel about the responsibility to define and live by your own moral code? Are you capable of doing so, or do you tend to simply justify whatever you want at any given stage in your life? Why do you think some people insist on a universal and absolute moral code?

There is much to like in religious moral codes, but there may be much to not like. Some people may ask what human impulse it was that created the Beatitudes and the inspiring stories of moral heroes, what natural instinct it is in humans that draws most people toward a natural social ethic, regardless of religious approval. Those persons can be encouraged to studiously develop their own moral code that builds on those proposed in the great faiths, using the best ideas from them, plus the best that they can add to them, and formulate a contemporary moral code better suited to an enlightened age.

They who know of no purer sources of Truth, who have traced up its stream
no higher, stand, and wisely stand, by the Bible and the Constitution, and

drink at it there with reverence and humility; but they who behold where it comes trickling into this lake or that pool, gird up their loins once more, and continue their pilgrimage toward its fountainhead.

—Henry David Thoreau,
"Civil Disobedience," 1849[9]

Claims of Insight into God's Will

How can anyone claim to know God's will? Many equally intelligent, pious, sincere people are positive that they know God's will with certainty, and yet their apparent messages from God are in conflict and mutually exclusive. Was there ever a war in which people of both sides did not claim to have God on their side? Who is to say which side is truly in touch with God and which just thinks it is? Is it possible that they are both entirely deluded? To credit a person or institution with knowing what to believe or do based on divine revelation requires one to conclude one has solid evidence to support all of several objective claims—each must be hurdled in order to claim that one knows what is God's will. With each step, the size of the crowd that continues to agree grows smaller:

- I know that God exists—despite what the atheists and agnostics have to say.
- And I know that He has consciousness—despite the deists and naturalists.
- And I know that He is aware of us—an assumption doubted by many.[10]
- And I know that He cares about our society—something not accepted by many pagans.
- And I know that He cares about what I do and think—again, the pagans disagree.
- And I know that He has many specific preferences for human conduct.
- And I know His exact stance on each contemporary issue is as follows. ...

Just the first step is an objective claim that requires proof; to then justify each of the subsequent claims requires yet more evidence. By the time one gets to the last claims of special insight into God's will on specific issues, one has quite a large burden of evidence to fulfill. No one is able to provide such evidence, and as a result, special claims of insight into God's will carry little weight in persuading rational persons to believe such claims.

You think that what I do is playing God, but you presume to know what God wants. Do you think that's not playing God?

> —Dr. Wilbur Larch, an abortionist,
> The Cider House Rules *by John Irving*

The Scripture Effect

Not all faiths, but some, hold that their scripture is the infallible word of omniscience, sufficient in its treatment of any issue addressed. Others may see such texts as the very fallible work of ordinary human beings who may have had much experience and better-than-average wisdom, but who could not have possibly considered all aspects of any issue, then or now. With the belief in infallibility comes the belief that the doctrine laid down in such scripture is immutable and timeless. We have already considered the advantages of holding such scripture as absolute. One disadvantage is that it teaches the mind to believe that once a policy or law is in place, even an entirely secular legal one, it is sufficient defense to say, "The law is right there. That's it. End of discussion." But, in fact, laws are human contrivances, and even the highest law of the land—for example, the Constitution of the United States of America—is subject to change by the people when they no longer feel that it meets their needs. There is a very specific and practicable "due process" for amending the Constitution, as there is for changing or repealing any law. As with claiming definitiveness by defending one's position "on faith," claiming definitiveness by defending one's position "with scripture" or with "the law is right there in the law books" is not the end of the discussion, but the beginning. The entire point of the challenge is to reform the law or the doctrine. Posturing as if that were unthinkable is not persuasive.

> *The people made the Constitution, and the people can unmake it. It is the creature of their own will, and lives only by their will.*
>
> —John Marshall (1755–1835),
> *a U.S. Supreme Court chief justice*

The Hippocratic Oath is a striking example of the Scripture Effect. It is an ancient text that is often highly, even solemnly, regarded in contemporary times when medical issues are considered. Oaths—or pledges—are serious business; individuals take very few true oaths in the course of their lives, and great social significance binds the person to any oath seriously taken. Many people outside of medicine, as well as in medicine, recognize the

Hippocratic Oath, if in name only, as not just a symbol of a physician's professional code of ethics, but as the ultimate reference for appropriate physician behavior. Surely a physician who knowingly violates the Hippocratic Oath is not a physician to be trusted; should the physician do so repeatedly and contrary to several of the oath's provisions, then one would hope the physician would be summoned to appear in front of the medical board and possibly also answer to lawyers.

Perhaps. Let us look at the exact words of the Hippocratic Oath. Please imagine that you are a freshly minted medical school graduate in the modern world and that you are expected to take this binding oath with formal ceremony and deep sincerity. Make sure that in your heart you are prepared to live up to each word of each sentence. Do you dare to pick and choose among its individual lines, deciding by your own conscience what you are willing to pledge, and what, by your own moral code today, you reject and perhaps strongly oppose?

I swear by Apollo Physician, by Asclepius, by Hygiaca, by Panacea, and by all the gods and goddesses, making them my witnesses, that I will carry out, to the best of my ability and judgment, this oath and this indenture. To hold my teacher in this art equal to my own parents; to make him partner in my livelihood; when he is in need of money to share mine with him; to consider his family as my own brothers, and to teach them this art, if they want to learn it, without fee or indenture; to impart precept, oral instruction, and all other instruction to my own sons, to the sons of my teacher, and to indentured pupils who have taken the physician's oath, but to nobody else. I will use treatment to help the sick according to my ability and judgment, but never with a view to injury or wrongdoing. Neither will I administer a poison to anybody when asked to do so, nor will I suggest such a course. Similarly, I will not give to a woman a pessary to cause abortion. But I will keep pure and holy both my life and my art. I will not use the knife, not even, verily, on sufferers from stone, but I will give place to such who are craftsmen therein. In whatsoever houses I enter, I will enter to help the sick, and I will abstain from all intentional wrongdoing and harm, especially from abusing the bodies of man or woman, bond or free. And whatsoever I shall see or hear in the course of my profession, as well as outside my profession in my intercourse with men, if it be what should not be published abroad, I will never divulge, holding such things to be holy secrets. Now, if I carry out this oath, and break it not, may I gain forever reputation among all men for my life and for my art; but if I transgress it and forswear myself, may the opposite befall me.

Many people may consider it wholly unacceptable to swear by pagan gods; to equate one's teacher with one's parents; to obligate oneself to give money away freely as a condition of their profession; to teach without compensation; or to discriminate (apparently) between sons and daughters. For some students, the sentence concerning abortion will seem equally absurd, though it will be embraced by others. On the other hand, the pledges concerning not doing harm knowingly and protecting the confidentiality of patients would be solemnly accepted by most modern medical students. All told, these lines come across exactly as one would expect the ideas of an ordinary person from a distant time and culture—some being relevant to today, and some not.

The striking point here is the difference between the reverence with which the Hippocratic Oath is often regarded (almost exclusively be persons who have never read it) and the more measured appreciation of the good and bad individual ideas and its overall ordinary human construction as it is typically regarded by those who have read it. Even some persons who have read it fully are so determined to keep this oath as their preferred source of guidance that, as they read, they apply whatever mental contortions, rationalizations, and "interpretations" necessary to render acceptable the unacceptable, somehow morphing the clear words of the text into the reader's preferred alternative understanding. When the words say one thing, they are capable of seeing in the words what they want to see. This effect of ancient writings being held as infallible and sufficient despite clear moral and sometimes physical absurdities is the Scripture Effect.

There is an alternative. People could neither embrace blindly nor reject blindly the ancient texts, but read them—those of their own heritage and those of very different heritage, those of the ancients and those of modern authors—and consider line by line what ideas are insightful, helpful, and otherwise worthy of adoption, and which are obsolete and perhaps harmful. Each person could choose to grant degrees of respect for each part of each text, without ever letting the feet leave the ground to call one text an infallible, timeless, universal, and complete code. The only problem is that doing so is very hard work.

Do not stifle the spirit. Do not despise prophesies. Test everything; retain what is good.

—The Bible, Thessalonians, 3:13

It is an unscrupulous intellect that does not pay to antiquity its due reverence.

—*Erasmus,* Works of Hilary, *1523*

Hero Worship

In contemporary society, we are often discouraged from making more than one God. Obligingly, we often do the next best thing: we recall our favorite personalities in family or in world history as the ideals we wish they had been. We may attribute to them greater virtue and little or no human weakness, thereby making them reliable, if fictional and unattainable, models of behavior. Stories and myths can be enormously effective, colorful, and entertaining teaching tools. It is difficult to overstate the human need for stories that inspire and teach. Archetypal plots, characters, and conflicts recur throughout the world in many forms to serve universal human needs; they serve as models for strength, courage, morals, justice, or any other great goal. The stories may be farcical, comic book–like, greatly venerated, or anything in between. Sometimes the characters are entirely fictional; often, they are fictionalized and heavily embellished versions of real persons. If the stories and the ideas that they teach are revered highly enough, the reverence is transferred from the lesson taught, or the "message," to a reverence for the hero himself or herself, apart from his or her message. The hero becomes the object of our reverence in addition to, and sometimes rather than, the ideas, choices, and behaviors taught.

However useful the stories, they should not be confused with factual history. We may risk forgetting to learn from both the good and the bad aspects of their lives; we are discouraged by our inability to achieve the same perfection. We would serve ourselves better to acknowledge in fair light the weaknesses that made them human as well as the strengths that made them extraordinary.

Reasonably intact histories of individuals, including our most honored, always reveal some imperfections and often some gross misjudgments.

Woman may be said to be an inferior man.

—*Aristotle,* The Politics

Negro equality! Fudge! How long, in the government of a God great enough to make and rule the universe, shall there continue knaves to vend, and fools to quip, so low a piece of demagogism as this.

—*Abraham Lincoln, 1859*[11]

The sex urge has been isolated from the desire for progeny, and it is said by the protagonists of the use of contraceptives that the conception is an accident to be prevented except when the parties desire to have children. I venture to suggest that this is a most dangerous doctrine to preach anywhere; much more so in a country like India, where the middle class male population has become imbecile through abuse of the creative function.

—Mohandas Gandhi, March 28, 1936[12]

$$G_{\mu v} - \lambda g = -\kappa(T_{\mu v} - \tfrac{1}{2}\, g_{\mu v} T)$$

—Albert Einstein, 1917[13]

Let us not allow this reality check to persuade us that there are no heroes; there are. They are real people who make extraordinary sacrifices for high ideals and save individuals and whole societies. Unlike the movies and ancient scriptures, however, they do not have bodies that defy physical law; they do not usually enjoy fame and adoration; and they are far from perfect. It is their very humanity that makes their choices and sacrifices heroic. They face dilemmas that have serious consequences, and they chose the hard or risky way instead of the easy or self-serving way. They are heroes for trying; they are heroes even before they succeed beyond their means and beyond the norm. It is their ordinariness coupled with their extraordinary choices that make them heroes. If the historical figures quoted above are indeed heroes, it is because they accomplished great things, humanity and errors and all.

The God Reflex can churn out not just gods, but saints, absolute laws, infallible scriptures, and unattainable models, which are dazzling but fictitious. They may all serve as inspiring goals, but the rational person will keep both feet on the ground and regard them for what they are, hopeful and man-made, mixtures of truth and error. If we choose to forget the humanity, mistakes, and blemishes of the things we idolize, and choose to readily accept fiction as fact, we distance ourselves somewhat from reality, we risk underestimating ourselves, and we may relinquish the power and will to make genuine improvements in this world to the "giants," the "gods," and the perfect persons who never were.

Thus far, we have looked at some of the lessons for rational thought as learned from history, science, and religion—all topics that engender strong passions. But what happens when our passions corrupt our speech? For this, we look at rhetoric—the tools of persuasion and of covert manipulation.

Chapter Five

Rhetoric: The Tools of Persuasion, Available for Hire

Rhetoric is the art of strategically using all the weaknesses, pitfalls, and fallacies of thought in one's own favor to win persuasion, often at the cost of the truth. People employ rhetoric when the genuine merits of their argument are insufficient to persuade. We must be aware of rhetoric being used to unduly influence us, and we must be equally aware of our own temptation to use it to persuade others illegitimately. Beware of being intimidated, bamboozled, lured, distracted, awed, humored, dazzled, railroaded, seduced, bribed, or otherwise pulled off the strict analysis of the merits of the argument—or caving in to the temptation to do so to other people—with techniques such as ...

Argument from Authority

Humans are primates. Primates usually have very strict social hierarchies. In other words, we respect authority, if we grant that the claim to authority is valid.

> *Anyone proved to be a seditious person is an outlaw before God and the emperor, and whoever is the first to put him to death does right and well. ... Therefore, let everyone who can, smite, slay, and stab, secretly or openly, remembering that nothing can be more poisonous, hurtful, or devilish, than a rebel.*
>
> —Martin Luther, church rebel—er, reformer, "Against the Robbing and Murdering Hordes of Peasants," pamphlet, May 1525

"It's true because ____ says so." Authority in a field can be real or pretended, but it never exempts that "authority" from rigorous scrutiny. The authority may be a person, a group, a revered institution, or a book. Many persons are so taken by authority that they don't even reconsider when the person admits that they are not an authority. For example, in an infamous ad in which an actor in a white coat is hawking a drug, he opens with, "I'm not a doctor, but I play one on TV." The ad, amazingly, was effective at convincing people to use the drug.

We must ask ourselves how the person or book received the status of an "authority." Through years of open scrutiny among other persons dedicated to the truth? Or was it more the "authority" that we grant blindly to traditional ideas, professional loudmouths, and manipulative or charismatic personalities? A Nobelist laureate *speaking on the issue of her expertise* does indeed carry more weight than the average person, but this authority stems from recognized processes of training, education, and collaboration with others of similar open scrutiny. But when she speaks on issues outside of her expertise, her authority may shrink considerably. Ask yourself, "Do I agree that the supposed authority is, in fact, an authority?" Do you believe her simply because she seems an authority? Does someone suggest that the person's authority immunizes him or her from any scrutiny?: "She is a famous scientist, so she couldn't be wrong"; "He is a pillar of the community, he couldn't have abused that child."

Christopher Cerf and Victor Navasky have written a book that is both funny and sobering, *The Experts Speak*, about authorities being completely wrong. History itself tells innumerable stories of the unknown amateurs making the breakthroughs. Authority matters, but it is not all that matters. We have a profound responsibility to fairly scrutinize authority.

> *Nothing strengthens authority so much as silence.*
> —*Charles de Gaulle, French president 1959–1969,*
> The Art of Living, *1940*

Taking Hostages: Coercing Your Decision

This is the rhetorical strategy of influencing what should be a rational choice by introducing a new and threatening consequence, i.e., coercion. Example: some persons with less money than they'd like have tried to reason politely with banks as to why the banks should just open the vaults and give them whatever they want. Usually, the balance of their argument, as perceived by a rational teller free to choose, is not persuasive. The client may then introduce elements of self-interest for the teller: flowers and friendship. The teller softens a bit and reconsiders his decision; the client's arguments may somehow seem a little more reasonable now. But then the teller realizes this is insufficient reason to waive the rules of the bank and declines. The client may then increase the self-interest for the teller: "I'll split it with you" (positive reinforcement, the proverbial carrot)—or, pulling out a gun: "If you don't, I'll blow your head off!" (negative reinforcement, the stick). Now the teller's judgment is skewed by compelling self-interest and fear, and the likelihood of a rational decision decreases (as for the money; whether it is rational that the teller will make it home for dinner is another matter). But the point is, clients such as these are not among the favored clients of the bank, because they are not discussing whether the teller should honor the request in a fair or rational way; the pros and cons philosophically do not remain the focus of the debate; the discussion is not geared for the enlightenment and personal development of either party; and the likelihood of a socially productive outcome diminishes.

This apparently silly dynamic of forcing a decision through a serious threat is actually quite common: "If you insist on dating that Jew, then you are not welcome in this home"; "You can choose to not believe, but then you will go to hell"; or "The CEO would like to remind all employees that negative statements about corporate policy are unhealthy to the corporation and may be inconsistent with a continued role on the corporate team." When we encounter such coercion, we are forced to either acquiesce or to openly acknowledge the attempt at coercion, reject it, and insist on returning to the merits of the argument.

But when is it coercion and when is it shrewd negotiation?: "I'll go to the store if you clean the bathroom"; "If you don't eat your peas, you can't watch television tonight"; or "You can't expect me to feel sexually intimate with you when I'm so frustrated about us not being able to go to Hawaii this year!" To discern the difference, we must ask ourselves:

- Why doesn't this person care about my agreeing based on the merits of the proposal, instead simply wanting me to acquiesce through either browbeating or bribes?
- Did something happen to trigger this new dynamic? Would solving that issue effectively solve this one too?
- Does the person trying to influence your decision have the right to do so?
- Which party is in the best position to make that decision anyway?
- Does the magnitude of the threat fit the magnitude of the request?
- Was there a previous understanding that is now being changed?
- Had you previously agreed that these two issues would be connected?
- If connecting these issues is new, is that truly unfair or is it just shrewd?
- Will the short- and long-term consequences, practical and emotional, of even making the proposition—not to mention of the consequences of the decision itself—warrant the benefits of the outcome in this one instance?

Issue Distraction

Your opponent in a debate shifts from the issue to something irrelevant: "You would vote for her for mayor?! She's a lesbian, you know"; "His brother is quite the town drunk, you know"; "His wife is into astrology"; or "I don't care to hear what Bertrand Russell says about ethics—he's an atheist, you know."

Perhaps the most common opportunity to hear politicians use issue distraction is when they are asked a question they do not want to answer. They give every appearance of answering, but don't actually answer the question. The skilled ones furrow their brow, give the questioner a sincere and thoughtful look, and, in tones earnest and reasonable, use more or less subtlety to wrench the question into the neighborhood of one of their own talking points that were scripted before the news conference. Immediately, the listener is trying to figure out how the answer connects to the question, is wondering what connection he himself obviously missed, is puzzling if the answer is really as senseless and irrelevant as it sounds, is gauging if he should interrupt, admit his confusion, and risk looking simpleminded in front of all his peers who seem to be following the answer just fine. Meanwhile, the politician is calling on another questioner.

It is worth pointing out here that in issue distraction, what one person considers irrelevant may be very different from what another person considers irrelevant. This is especially true when the issue is distracted to a matter of the person's character. That brings us to. ...

Ad Hominem Attack

"To the man." This is a specific form of issue distraction in which the issue is shifted to an irrelevant but unflattering aspect of the opponent. You disagree on the subject of national defense spending? Then accuse your opponent of using drugs, having an affair, or mismanaging corporate accounts. Don't see eye to eye on health care? Then speculate on things entirely of your imagination: that your opponent could only take such an insensible position if he were scarred from a past emotional experience or if he had ulterior motives, a hidden agenda, which should make any thoughtful voter nervous. Anything to knock him off balance except the merit of your evidence on the original issue. Even the least informed voter fancies himself a good judge of character, so a practiced Machiavellian will try to destroy his opponent's general reputation.

Enough facetiousness. Surely "character" is a critical ingredient in nearly any role of trust—but there are also many other critical ingredients, such as experience, knowledge, and skill; and there are no perfect people in the world, so there will almost always be some type of skeleton in the closet. (Okay, maybe not with Michael Jordan, but with everyone else.) "Worthy of admiration for writing the U.S. Declaration of Independence? In 1826, he still had slaves at Monticello!"; "You'd let her be head of the EPA when she 'messed up' her taxes a few years back?"; "I can't see letting a man be police chief when he got a DUI in college"; " ... he had a sexual dalliance, ... "; " ... he did a line of cocaine in high school, ... "; "... he was a philanderer, ... "; and so on.

So the questions become, Is this enough information to rightfully judge the person's character? Do we have sufficient evidence of all the person's decisions through life? Is this one event, if true, and in the context of their times, enough to outweigh the person's more practical and relevant merits of experience, knowledge, and skill to tip the scales significantly?

Is it possible to give people credit for their merits and achievements, hold them accountable for their faults and mistakes, and remain focused on the relevant issue? Perhaps for many voters it is not. One reason why ad hominem attacks work well in a democracy is that many of the voters know little about the person's experience, knowledge, or skill, so the salacious story is all they know, and the scales tip accordingly.

It is not an ad hominem attack to justifiably scrutinize the relevant credentials or the relevant disqualifications of a person. In politics, some "attack ads" are, in fact, perfectly valid criticisms of relevant and embarrassing events in the candidate's history of public service. True ad hominem attacks, however, are a type of diversion from the issue at hand, and, as with other types of rhetorical diversions, we must calmly point out that we are aware of that attempt at diversion and wish to resume discussion of the relevant issues at hand.

Dehumanize the Opponent
One very special form of ad hominem attack deserves particular attention: simply and sweepingly defining the person as being irretrievably beyond reason. Reason is the defining characteristic of our species (even if we all need to polish it a little), and to declare a person incapable of reason is to wash away any value to his opinion. One may denounce the opponent as insane, hysterical, or—better yet—evil. This is especially useful in political debates, personal or international, when a person is trying to persuade others that there is no sense in trying to understand the other side of the debate; that indeed there is no understanding it no matter what attempt at discussion is

made; no more time should be wasted before something rather one-sided is done; and anyone whose opinion is in any way sympathetic to evil must be himself very muddled or evil. This also handily reduces the responsibility of the speaker to give his or her own reasons as, presumably, any at all would outweigh the total deficiency of reason on the part of the other.

Once we are all in agreement that the opponent is beyond reason, then we need have no reservations about his or her very different set of facts or perspectives, reasoning, or responses. The reason this ploy is illegitimate is that few people are truly insane (a medical diagnosis that is inconsistent with being able to plan and execute complex logistical and socially integrated activities, such as acts of war), or hysterical,[1] or evil (especially if "evil" is intended to mean without any conscience or moral restraint). Persons genuinely out of touch with reality are quite unlikely to be able to hold a job, attend meetings, organize arguments, or direct armed attacks. A person capable of such things is not insane, but just differs from us utterly in his set of facts (accurate or just perceived to be accurate), perceptions on the issues, notions of reasonable solutions, and assessments of moral propriety. Those persons do have reasons, just reasons that seem so different from our own that we did not even consider them prior to this conflict; thus, it is not necessarily *his* intellectual deficiency, but perhaps our own that is the problem.

Although a person may not be "evil," he may still fully deserve to be dealt with quickly and severely. He may not really be evil, but he may still need to be crushed as a bug so he will hurt no one else. A very dangerous, recurrently aggressive person may necessitate the most extreme action. Yet *having* to react thus begs why. Was the entire population of 1930s and 1940s Germany, or even just the Nazis within it, all insane and beyond human reason? Or were they normal people (normal including some who would be xenophobic, demagogic, and violent) who were resentful from decades of economic austerity imposed by the reparations of the Treaty of Versailles? As justified as we feel (and may actually be) in fighting back with all possible severity, the entire situation represents a failure of reason at some point past—theirs or ours—and our swift retaliation must still be thoughtful and not a continuation of such failures.

Evil may be largely in the eye of the beholder. The other person's gross moral violations are evil, whereas ours are in some way regrettable but excusable consequences of human nature and the hard facts of life. Chechen rebels can conduct a hostage scheme that results in the murder of hundreds of ordinary school kids, teachers, and parents, and this may be another example that comes as close to "evil" as anything could to deserve the term.[2]

So, too, with terrorists flying fully occupied commercial airplanes into buildings full of civilians. These persons may say the same about the United States being a source for unrelenting pornography, automatic weapons, global climate destabilization, and exploitative trade policies that impact the lives of millions. "Evil" may be a bit too subjective, as well as clinically inaccurate, to be useful in reasonable debate.

If the person who did some heinous act is not truly evil, then what else could possibly explain it? Ask him. Listen for a long time. In addition, it may be a revealing exercise to imagine that the supposedly evil person is actually an ordinary person who is as rational and fair minded as you are. If he is, then what would it possibly take, historically or personally, to cause such a person—you, for example—to turn to the committing of atrocities? You would have to feel unquestionably superior or unimaginably violated, unheard, out of time and alternatives, and certain that your values are righteous, all of which may seem to you to justify "cracking a few eggs." Such is the path in the minds of the aggressors leading to the extermination of Native Americans, Kurds in Iraq, Africans in western Sudan, and species and ecosystems throughout the world, to name but a few of history's tragic tales. The path to justification is similar in every case: *our* acts in history are understandable, but those of the *others* are evil. This is a nearly universal belief and fallacy.

Thus, to call a person or policy insane (in the rhetorical, nonclinical sense) or evil is less a declaration of the opponent's inhumanity than it is a declaration of our utter arrogance and of our utter ignorance of his facts and viewpoints. If we seek to stay rational and to learn the real reasons for the act, we may have to address our arrogance and our ignorance, but doing this requires laborious discussion and meeting of the minds, and that is often much less satisfying than swift, draconian, and morally unhesitant action that avoids the possibility that, at least on some issues, our opponent may be right.

"You Are Beyond Help."

This device is just a cheap shot delivered with retreat—it at once jabs an insulting condescension at the other party while abandoning explanation (usually because the inadequacy of the explanation is becoming obvious): "Never mind—you just don't understand"; "You just can't grasp the science of palm reading, so I won't try to explain it to you"; "You think herbs don't work because you refuse to listen to the evidence"; "Some people are tuned in to the presence of the spirits and some just are not"; or "I know it may seem unfair, but you just don't understand how the system works." Again, nothing has been proven either way. This is an attempt to escape the discussion without admitting weakness in one's view.

Appeal to Pity

This is a form of distraction from the issue with the claim that the logical outcome would be painful. The appeals are often non sequiturs: "You can't give me a C on this paper because then I might not get into medical school!" or "I don't think that my taxes this year are fair because there won't be enough money left for me to take care of my mother." These arguments, while indeed pitiful, are often not pitiful in the way the speaker intended.

Ad Ignorantum

"Appeal to ignorance." This is the specious argument that a claim is true or even reasonable just because it has not been shown to be false or because an alternative is not 100 percent certain. It takes advantage of a lack of perfect data and of those persons who do not know how to think probabilistically. If Explanation A is 99 percent certain and lacks only the most theoretical definitions of certainty, its deficiency is not just cause to declare the preferred and alluring Explanation B, which seems less than 1 percent likely, to be a reasonable alternative explanation. We often do not have certainty, but relative measures of plausibility can be compelling. The difficulty is that most people are uncomfortable with weighing degrees of likelihood and prefer certainty regardless. Thus, any degree of uncertainty for the unwanted answer is an excuse to reject it in favor of the preferred, though less likely, explanation.

The example above is an extreme one to illustrate the point—most issues fall in the middle, with Explanation A having maybe 75 percent likelihood and B having maybe 25 percent likelihood. Being able to discern fairly the degree of evidence for and against each alternative is a critical skill. But claiming that "until the other idea is proven 100 percent, it is reasonable to believe anything I want to" is irrational.

Whether A is true or not may be mutually exclusive of whether B is true or not. In those cases, evidence for A may be an argument against B. The more evidence we have that Nicolaus put the worm in Rachel's shoe, the less likely it is that Alex did it. If it's 75 percent likely Nicolaus did it, then it is only 25 percent likely it was Alex; as evidence under Nicolaus's fingernails pushes 75 percent to 90 percent, then 25 percent shrinks to 10 percent.

Clutching to a few shreds of hopeful and weak evidence while waving off a much greater and more valid (if imperfect) collection of opposing evidence is a sad exercise of forcing the scales to tip as you wish they would; it is a last-gasp delusion in an argument you refuse to concede. When an idea is less well supported than another or when it has been disproved, be courageous enough to admit it. Going down with the ship may be considered noble among mariners, but it is foolhardy in personal deliberations and public debates, and it only delays progress.

"I say a huge beast lives in the center of the Earth. We don't have instruments to check; therefore, you haven't disproven it; therefore, it is reasonable for me to say with confidence it is true." In this case, the argument is unpersuasive because there is some compelling evidence to consider—for example, that we know the center of the Earth is too hot for any complex biological organism.

What has been attempted, however, is to dodge and shift the burden of proof. Ideally, the burden of proof is shared by all, and those with the least plausible idea bear it more so. However, as per the "Obligation of the Underdog," in daily life, where hard data is often lacking, the status quo remains the default conclusion of the

> Imperfect data begs for better data, but it may still be compelling as it is.
>
> For any question, if a person discovers a troubling lack of adequate data to give an answer with confidence, it would be an enormous contribution to dedicate oneself to doing new and original research. This may bring significantly better data to light for the consideration of all. While it is great fun to collect *existing* data and analyze it oneself, that endeavor depends entirely on those committed and tireless souls who may have spent years in obscurity to produce the gems of data over which we pour and from which we all benefit. For elevating the level of the debate, we are in their debt. We begin to repay them by analyzing this and related data in a fair and rational manner; the greatest repayment is to build onto their work with yet more original research. When we "stand on the shoulders of giants," most of us are enjoying the view, but some are actually bringing the effort to greater heights. Hats off to them.

masses, whether it deserves to be or not. For example, people may choose to believe something because it is comforting (the afterlife) or exciting (alien abductions) because they can always claim that *absolutely* disproving them as possibilities is virtually impossible. For the masses, the burden of proof lies with the one who would change the status quo.

Human nature tempts us to take advantage of the lack of data to do whatever serves our desires. Example: a physician who is very close to me was very discouraged one day. He explained that in his large group practice he had observed an increasing haste in seeing patients, driven by administrative policies that were, in turn, driven by ever increasing salary expectations that had already put the physicians' salaries well above that of their physician peers. He had approached the administrators to express his concern that while the high salaries were satisfying, he was worried about the impact this method was having on the quality of assessment and treatment of the patients. The physician stated that he wanted to institute a quality-assessment system in which charts would be randomly reviewed. He expressed his willingness to make less money if the chart review showed that the doctors needed to slow down. He was told that there was no need to set up a formal data collection system on quality of care because, as of yet, there was no formal data on hand to suggest that there was a problem, only a series of observations. Furthermore, he was disallowed from establishing any such data and quality monitoring system.

Ad Populum

In all very numerous assemblies, of whatever characters composed, passion never fails to wrest the scepter from reason. Had every Athenian citizen been a Socrates, every Athenian assembly would still have been a mob.
—*James Madison, in* The Federalist Papers, *No. 55*

"To the mob." This is appealing to the emotions of the crowd or looking to agreement by the ignorant crowd as validation. The size of the crowd may be impressive, but their expertise and thoughtfulness may not be: "Come on, everybody does it." or "Are the people of this great country going to just roll over and let this international treaty insult us?" It may be a sobering commentary on general education that *ad populum* appeals are characteristically oriented toward base or violent impulses. What may not persuade individuals can often be used to incite millions. An illusion of consensus is created that is persuasive to the ignorant and intimidating to the pensive.

... even if men went mad all after the same fashion, they might agree one with another well enough.
—*Francis Bacon (1561–1626),* Novum Organum

There are times when public opinion is the worst of all opinions.
—*Sebastian-Roch Nicholas de Chamfort, French writer,*
Maxims and Considerations, *1796*

Just the suggestion that "most people" agree with an idea can stifle a challenge to it. A common ploy is to shield a comment from scrutiny by leading with "It is generally accepted now that, ... "; "Few people really doubt anymore that, ... "; or "Most thoughtful and informed people realize that, ... ". The claim that an unassailable number of people would oppose your disagreement is meant to cause you to feel foolish in voicing any doubt. Whether such an imbalance of opinion exists may or may not be true; even if it were true, it does not necessarily support the validity of the idea.

This tendency of human nature to assume as valid whatever consensus holds has unsettling implications in a time of globalization. More than just hamburger restaurants, bananas, chickens, language, and economic systems are spreading and monopolizing, gradually replacing an array of completely different varieties with uniform monoculture. Fundamental assumptions about human existence—our purpose, our ultimate goals, our most central institutions, our accepted modes of behavior, our "appropriate" priorities and values in life—are also spreading. With the loss of visible examples of fundamentally different societies and ideas, new generations may make the *ad populum* error that the ideas held by consensus are more valid than the ideas vanishing or extinct. The dominating ideas may have many reasons for rising to dominance, such as power, saturation, or seductive appeal to the masses, but none of these are the same as having a unique claim to being socially "best" or factually accurate.

But let us end this section on a more optimistic note. Many times in history it has been the people who have thrust progress upon their leaders and tyrants.

> *Greater than the tread of mighty armies is an idea whose time has come.*
> —*Victor Hugo (1802–1885), French writer*

Ad Lupi (To the Wolves)

Sometimes, rather than calling on the ignorant crowd to support you, you may first see a popular groundswell building against your opponent for some irrelevant and base issue, and you can choose to just quietly step aside while the crowd consumes him. When they disperse, the floor will be yours alone.

Suppose you are involved in public policy or in the intrigue within a corporation, and there is to be an open debate and a definitive decision on serious policy issues. However, just prior to the debate, your opponent is accused of sexual indiscretion in her personal life, poor judgment in remote days of youth, or some other vulgarity that would surely inflame the populace or the staff and remove her from office and debate. Do you stand aside and let it happen? Do you enthusiastically fan the flames?

"Well, So Do You"

In Latin, *tu quoque* (pronounced *tu kwo kway*). Similar to the *ad populum* argument, this is another type of appeal to bad example—in this case, your opponent himself: "Well, so do you." (Perhaps we should call this, alternatively, the Na na na na na Argument.) If someone accuses you of cheating on your husband and you respond this way, nothing has been proven or even argued either way. Whether or not your opponent is a hypocrite is a separate issue. However, learning your opponent's reason or justification may give insights for both of you on your own reasons, as well giving you the satisfaction of turning the tables. In the end, both parties are usually not being strictly rational or consistent. This is unlikely to be a particularly lofty discussion.

Poisoning the Well

Would you take a drink of water from a well that people say is contaminated?

In ancient Athens, the great leader Pericles faced a demagogic rival, Cleon. When substantive debate on issues gained Cleon no ground, he instead attacked Pericles's close friend Anaxagoras, who had described the sun as a mass of stone on fire rather than as the god Helios, which the general population thought it to be. Pericles defended him, but Anaxagoras was convicted of impiety and the character of Pericles fell under suspicion.

In contemporary times, there are many examples of politicians and their agendas being derailed by popular outrage at sexual indiscretion, family embarrassment, or salacious revelations. Whether these are rational and efficient ways of choosing our policy makers and policies is the question. If your child were desperately injured, would you refuse the best surgeon because his personal life was spotty? In a world of serious and complex policy problems, is it in our interest to allow any thoughtful policy maker, whether he agrees with us or not, to be removed from the debate due to unrelated personal issues? Do you expect your policy leaders to be both supremely competent in their profession and personally flawless? How much are you willing to sacrifice in professional standards to meet your personal ones, and vice versa? Until we have perfect candidates, we will have to make tough choices; this can be done with more reason or with less.

Would you take advice from a source who has a reputation for misleading people? If you can portray the source of an argument as suspect, then his or her argument or ideas may never be given a chance: "The girl is known to be a poor student—and you believed her?"; "This article is biased and poorly researched. Let me know what you think of it"; "So my daughter goes to see this doctor in a crumbling old office building in a bad neighborhood and he tells her she has … " Notice that the way in which the person is portrayed in a negative light may not necessarily be valid evidence of a character defect; even if it were, it may not be relevant to the issue at hand. It may be hard to judge when indeed the well may be poisoned. Would you trust the eyewitness account of a man whose wife was charged with theft? Would pointing out their relationship be "poisoning the well," or is it a valid reason to discount him as an objective source? Poisoning the well can be an effective, if underhanded, way to see that your opponent's idea is never fairly considered.

Tautology

This is a self-justification by circular argument: defending an idea with the idea itself so that no further information is given despite the appearance of having done so: "That's just the way it is"; "Because"; "That's just the way corporate America does things"; "I believe B because I already know A is true; and I believe A because I have already proven B"; "I believe this scripture is infallible because it is the word of God; I believe in God because the infallible scripture says I should"; "I believe the results of this study are scientifically valid because it was published in a trusted journal; I believe this journal is trustworthy because it publishes valid studies"; "I am because I am"; "Marijuana must be illegal if it is dangerous; it must be dangerous if it is illegal."

A tautology that is so common that it deserves special mention as a fallacy unto itself is the tendency to dismiss an idea because of its source, which you dislike because of the nature of the ideas it typically produces. For example, "I find that notion about religion invalid because it comes from a person whom I find irritating and offensive; I have always found that person irritating because he has produced invalid notions about religion." You need instead a valid reason to reject the idea based on its merits apart from its source or a valid reason to reject the source based on its merits apart from the ideas produced.

A classic tautology made the headlines when President George Bush was asked why he said there was a relationship between al-Qaeda and Iraq when an interim report of a bipartisan panel investigating the decision had concluded that no evidence had been found to substantiate that claim. His response was, "The reason that I keep insisting that … there was a relationship between Iraq and Saddam and al-Qaeda because [sic] there *was* a relationship between Iraq and al-Qaeda" [the emphasis was his].[3] The point of the question was, of course, to hear the evidence that Bush often referred to, but, to date, had not revealed. The evasion was a tautology.

Unproven Assumptions

"We must establish a curfew to reduce teen crime." Has it been shown that indeed a curfew will have the intended result? Recall the explanation of flawed syllogisms: identifying the underlying assumptions, or "premises," is often a very helpful way of discovering why two parties are disagreeing. "We were arguing over and over about whether or not we should spend the weekend at the beach, until we realized that he assumed that by spending money to go to the beach we would not be able to go to his mother's for Christmas as planned, whereas I assumed we would see his mother regardless. So we looked over our budget and realized we could do both."

False Dichotomy

Issues are not necessarily either/or, and life is seldom true/false. It's much more often multiple choice of the worst kind: sometimes the answer is all of the above; sometimes it is some of the above. We tend to forget the possibility of an all-inclusive explanation or the some-of-each middle ground: "So what determines behavior—nature or nurture?"; "If you are not with us, you are against us"; or "We can't support this missile program because we need the money for schools." The important skill is to be able to notice the assumption of a dichotomy and ask if it is necessarily so: is there room for a compromise or a third option?

Bundled Ideas

If someone tries to tell you, explicitly or implicitly, that you must accept or reject an entire set of ideas as a unit, you can reject that proposition. It is an all-or-nothing wager that doesn't have to be—a form of a false dichotomy: "How can he claim to be a Republican when he voted to increase taxes for the schools?!" or "Do you believe in the ideals of the Revolution—yes or no?" Such a set may be the various ideas espoused by a legislative bill, an organized group, or the teachings of a single person, for example. Your opponent wants you to buy into the whole set of ideas with no discussion of the individual points. It simply is not a logical proposition, because bills, groups, and individuals are far too complex, representing far too many ideas and acts. Ideas may tend to come to us bundled, but we always have the freedom to intellectually "unbundle" them for individual consideration—in fact, it may be our responsibility to do so.

I think one should do in philosophy what customarily happens in the Senate;
when someone proposes something which pleases me in part, I tell him to

divide his proposal into two parts, and I support what I approve.
—*Seneca (4 B.C.E.–A.D. 65), Roman statesman*
and philosopher, Letters on Morals *(21.79)*

Some people will find the *un*bundling of their ideas greatly annoying, perhaps because it is more taxing to defend each idea individually; perhaps because they were consciously trying to sneak in the less popular ideas with the more popular ones; or perhaps because it renders convenient labels of individuals terribly inadequate and inaccurate: the welfare appropriations bill with the line item about building a new highway in Toledo, for example. In the Catholic Church, members who choose to accept some doctrine and to personally reject others may be derisively if somewhat humerously called a "cafeteria catholic".

It is in human nature to be averse to complexity, so we lump things together and pass judgment on them as a whole. We like being able to say that we like or don't like this person or that group. We don't want to be bothered with the details within. If we express some mixture of honest agreement and disagreement with a large group—not to mention doing so for both opposing groups—we will find ourselves distrusted by nearly everyone, except perhaps by the intellectuals. Some may find this a favorable trade.

It is possible to analyze the speaker or the organization one issue at a time. Despite pressure otherwise, one can choose to accept some of the ideas and reject others.

Labels for groups may offer a very general description, yet they are often very inaccurate for the individual or for sizeable portions of the group: "Muslims are fanatic"; "Republicans are financially irresponsible"; or "Democrats can't make a decision." Reason, accuracy, and justice dictate that we reject the use of simple labels.

A brief note about the legislative implications for bundling and unbundling ideas: bills are themselves bundles of ideas, and it is impracticable for every line and every idea to be analyzed and voted on separately, even if that were ideal. Thus, it is delegated to committees to analyze each idea separately. Perhaps a goal could be to reduce the size of each bundle as much as possible. Regardless, it is evident that some legislators manipulate the bundling to include a completely unrelated provision that would not, by its own merits, pass. One justification is that it "should" pass, but manipulative politics prevent its passing, so now justice rides in on the back of the garbage truck. Perhaps. Another justification is that everyone in Congress has a pet project ("pork"), so if we put them all in, then everyone is served. But the process does involve actually scrutinizing the merit of the individual ideas.

The One-Cause Dodge

We must resist the temptation to simplify what is not simple.

—E. R. Dodds

Issues are often complex, with multiple influencing factors. But most people have a hard time getting their hands around multivariable complexities, so they try to pare them down to an argument about just one cause: "That's not the cause of the high violence rate in society—financial desperation is!"; "No, it's violent video games!"; "No, it's absent parents!"; "No, it's drug use!"; "Too much religion?"; "Not enough religion?"; "Violent movies?"; "People unpracticed in self-control?"; "Glorification of weapons?"; "Lax gun laws?"; or "Racism?". Of course, there may be a valid argument for each and any of these, but by pointing the discussion toward deciding which is the "main" cause, no one cause will ever seem sufficient to explain the problem, so there is only paralysis.

A fanciful example may help.

Suppose I feel unsafe walking in my yard because I have an intimidating swarm of bees that have stung neighbors, some quite badly, and that have intentionally scared me with their buzz-bys, not to mention their unnerving glares. I feel as though they are watching me, wondering how much nectar they could get out of me.

Suppose that I am greatly annoyed, and I want my yard and my neighborhood back. I want the bees gone—most of them anyway. I complain to the bees ... but the bees say they're just going where the flowers are. The flowers begin pointing fingers: the columbines would blame the larkspur, the larkspur would blame the chickweed, the chickweed would blame the spring beauties ... down the line to the two dozen species of flowers that occur in my yard. The flowers as a whole would claim the right to make a living and blame the birds as a whole for their lax regulation. The birds, no doubt, would blame me for inadequate support at their watering dish, and that I thus make their job that much more dangerous, not to mention underappreciated. Meanwhile, the bees would happily buzz about their frightening business, shutting down much of the healthy work that needs to be done in the yard. I would remain cowering in my home, just as tormented as ever.

Instead of allowing everyone to argue about which single cause is to blame, I first get the facts—how much time is spent at which flowers, how much water I can give the birds, and so on. Then I make a plan to reduce and displace bees by addressing every reasonable factor.

My assessment and plan: every factor contributes to the problem a little bit, so every factor makes a sensible change: 20 percent fewer columbine, 20 percent fewer larkspur, 50 percent fewer chickweed … There will be some unavoidable injustice to some of the flowers, granted. I'll try to minimize it, but won't let myself be paralyzed by imperfect data. Bring more water to support the birds and to attract more birds. And the most bothersome bees should be dealt with one-on-one. I'll carefully monitor the results so I can modify my approach next spring.

A real-world example: I was a guest at our state capitol to discuss health-care reform issues when I stepped onto a crowded elevator. A senator asked me, in a loud voice, "Dr. Hindes, in the few seconds we have before we get to the third floor, what is the *one thing* that we must do to solve our health-care problems?"

All eyes were on me. "Senator, the *one thing* we must do is get past the simplistic notion that there is *one thing* that we must do."

Ding! Saved by the bell.

Don't let progress come to a standstill by arguing over which single factor bears the most responsibility. Don't jump to simplistic conclusions. Get the facts. Expect that complex problems will require complex solutions. Don't shy away from the multivariable complexities, irreducible uncertainties, or unavoidable injustices. Get the best data you can, do everything that seems reasonable, learn from the experience, and do a better job the next go-round.

> *For every problem there is a solution that is simple, direct, and wrong.*
> —H. L. Menken, *as quoted by*
> Graham Hancock *in* Lords of Poverty

The Myth of the Slippery Slope

"Dominophobia": "If we outlaw abortion at nine months, then we'll lose the right to abortion at all"; "If we outlaw automatic weapons, then they'll take all the guns away"; or "If we allow gays to marry, we'll have to legalize polygamy."

Take a step back: is there really a slippery slope at all? Just because there is a spectrum of related situations here, are there really no distinct incremental changes as we move down the spectrum? Might it be less of a slippery slope and more of a level park bench on which we can choose to sit anywhere and just stay there, if we want to, with legitimate, if imperfect, rationale? When the Americans lost the Vietnam War to the communists, there was no cascade of other Asian nations turning communist as had been predicted, because the circumstances in each nation were sufficiently different to prevent any "slipping."

Even if the spectrum were a bit sloped and a bit slippery, are we not capable of weighing all considerations and drawing lines through grey areas when we need to, despite the inevitable inconsistencies? This may be imperfect, but it is still more satisfying than collapsing into one extreme or the other. You could always take responsibility for not slipping and for making only reasoned, deliberate decisions.

People who are uncomfortable with complexity, with balancing multiple conflicting considerations, and with negotiation will become alarmist and declare any compromise a total loss and an abandonment to the inevitable slide of the slippery slope. They underestimate the power and responsibility of people to use reason.

We must dedicate ourselves to rationally finding the "middle way" between the conflicting interests at the two ends of any spectrum. We are capable of rationally holding our position, even if the slope is a bit slippery, until reason tells us to change to a new, firm position. We can courageously face the grey area, make tough choices, accept the inevitable inconsistencies, and give good reasons for our position.

The Straw-Man Argument

Ad absurdum, taking it to the absurd: "I can't believe you would criticize this one line in the whole book—I guess you find the whole book worthless." In this related ruse, a person may make no explicit claim at all to a slippery slope, but repositions your idea in an extreme and untenable form and attacks that instead, leaving the impression that the original statement has been proven unsound: "The senator from Ohio is acting as if this little bill of mine is going to be the financial ruin of the nation, but, as my data shows here, the nation will not collapse if we build one more airport."

The phrase comes from the notion that it is much easier to beat up a challenger if he is a straw dummy made and set into position by you rather than the actual flesh-and-blood opponent: "Do I think it is okay for you to study with that guy from your class? Since when are we dating other people? I had no idea we had an open relationship and could just sleep around!" or "You want a puppy? You already have a goldfish. This place isn't a zoo, you know." Just because an idea taken to its extreme may be unjust or absurd does not mean that the original idea, applied within reason, is invalid.

Persons who ask for evidence—any evidence at all—frequently encounter this device:

> "That's an interesting assertion, Governor. Do you have any evidence to support it?"
>
> "Well, if you are asking for *absolute proof*, I guess I don't have that, but my point still stands. ... "

Society offers plenty of extremists to be the straw man for us. It is a common tactic in less civil debate to cite an extremist from one's opposition group and insist they represent the entire religion or political party, and so on, as condoning the same ideas or behaviors. So the Republicans are the party of apocalyptic, snake-handling evangelicals and the Democrats the party of seditious communists who would nationalize all industries, according to some. Admitting the breadth of opinion, the moderate majority, and the reasonable arguments on both sides is difficult and far less gratifying to a debater's bloodlust. The world would be simpler and our crusade better justified if everyone "on the other side" were an extremist; the problem is, they're usually not. The zealots may make better headlines and more memorable stories, but usually the great majority of persons and events are much more reasonable than we'd like our enemies to be.

Intimidating Certainty

Present an idea as beyond debate, already decided, and quickly move on before anyone calls you on it. "Everyone knows it would be impossible to raise the age at which people can collect Social Security, so let's look instead at what corporate taxes we'll have to increase." or "Since we all know that we've cut the budget at the corporate headquarters to bare bones, let's look instead at the lavish benefits plan that the janitors are getting."

The very force of the apparent certainty raises suspicion about the actual certainty.

What is earnest is not always true; on the contrary, error is often more earnest than truth.

—*Benjamin Disraeli, British prime minister,*
in a letter to Queen Victoria, November 4, 1868

Hyperbole

This is deliberate and obvious exaggeration, used for effect: "This bill would be the ruin of the Republic" or "This proposal on new zoning laws is the greatest advancement for justice since the Emancipation Proclamation."

Drama

"I've sent our brave men in to rescue those poor souls huddled on their rooftops. ... This is no time to discuss who voted for what on development in the floodplain. ... Now we can show you this astonishing video clip of a terrified family dog being rescued by our courageous fireman. ... "

Amuse and Deflect

Sarcasm, humor, and charm are all forms of distraction that can be so amusing (although perhaps not to everyone) that the point in question is momentarily evaded. Usually, the subject is then deftly changed before that point is again addressed in serious fashion. In the least, the person who drags the conversation back to the difficult point is seen as a stick-in-the-mud who is not allowing the others to enjoy the flow of the conversation; he must now face the disadvantage of annoying the group with seriousness of purpose.

False Analogy

Consider whether there are enough similarities to warrant the analogy or if the speaker may be trying to spin the issue somehow: "You should have seen him leaving the podium with his tail between his legs, ... "; "This company is raping the environment, ... "; "You know, my candidate reminds me a lot of John Kennedy, ... "; "Looks like a real war is building up between these two laundry-detergent makers, ... "; or "She's turning the PTA into a fascist state!".

Putting Words in Your Mouth

This is usually done in an effort to turn your argument or suggestion into a straw man, which is more easily discounted or disproved: "Yes, yes, some Puritans boiled some Native Americans alive for heresy—so then you are saying that all religion is evil?"; "Can't come over for dinner?! If you are saying you are breaking up with me, then just do it"; or "Opponents will tell you that this new little tax will drive a stake through the heart of every honest businessperson, but there are many reasons to believe it will help."

Assumption of Persecution

Perhaps a special form of the fallacy of putting words into your opponent's mouth is the common device of putting into your opponent's mind a malicious and prejudicial intent. Thus, for example, when an employee is reprimanded for being habitually late, he may choose to believe that he is being harassed because he is black. Persons may object to Israeli policy with the same vehemence and reason that anyone may object to any governmental policy, but in this case, they are charged with anti-Semitism. If an American is critical of a specific American policy, he is charged with being generally unpatriotic; of a Christian doctrine, he is anti-Christian; of woman's issue, he is misogynistic. This is the pervasive and toxic "identity politics" that people hide behind today to avoid rational critique.

The reason for claiming persecution is that it relieves people of the possibility that their critics are right, of the responsibility of genuine reflection, and of the hard work and humble pie of reform. It shifts the spotlight off of my inadequacy and puts it on the much more grave moral failing of my critics.

In any one case, the persecution may or may not be real, but it depends on the merit of the criticism and on the consistency and fairness of the scrutiny, not on the final position taken. To look for genuine prejudice, one must assess the reasonableness of the individual arguments and look for fair concessions or admissions of understanding. To acknowledge one's own sensitivity on an identity issue and consult persons known to be less sensitive, more moderate, and unprejudiced, and take seriously their interpretation, may keep one's reaction balanced.

Semantics

Being clear about the definitions of the terms discussed is essential, but people can also hide behind disingenuous disputes of common definitions: "It depends on what the meaning of 'is' is."[4] Along the roadside of every debate is an entangling jungle of semantics that suddenly seems full of dissectible curiosities for those who have realized that the road is not coming out where they wanted it to. But, to be fair, some people don't flee into the thicket; they just get lost and forget where they were going. This was the case even for Aristotle, according to Durant: "He longs to think clearly, though he seldom, in his extant works, succeeds; he spends half his time defining his terms, and then feels that he has solved the problem."[5]

Defining one's term is essential to clear communication, but do so and get on with your point.

More often than persons intentionally bogging us down with undue analysis of definitions, we ourselves are to blame for listening too carefully to the literal words chosen, rather than to the person's meaning. Unless you are a lawyer at work, give the person the benefit of the doubt: you know his intent; you know the general direction of the argument; you know that he is not trying to trick you—quit pinning him down on particular figures of speech, let it go when he uses an indelicate choice of words, and let the man make his point. With practice, you may develop an automatic translator in your head that can avoid unnecessary quibbling and bridge huge gulfs.

Hidden Agendas

It's a fact of life that sometimes the person is not being truthful because they have another intent than the truth—to make money from you ("My product works better than my competitors"), to gratify you ("He'll get what's coming to him in the next life"), to prop up his own self-esteem (Why hate your own inadequacies and labor to improve yourself when you can just go beat up some black people who are doing better than you are?), or to offer an answer for the satisfaction of being able to offer an answer (Every society had great creation myths and some half-baked scientific theories). These ideas and theories may seem far from good science, or even justice, but they offer something else that is compelling and attractive, so rationality takes a backseat.

"Hidden agendas" are not necessarily insidious or malign; many people simply aren't ready to admit to themselves or others what they really get out of their belief. If it hurts no one and even confers some benefit, why argue? Consider the five-year-old child who will soon die of cancer. She is very upset to hear someone say that Santa Claus doesn't really exist, and she asks you. What would you say? Under what circumstances would this be analogous to arguing with an adult who derives great pleasure from her religion, when you feel it is benign although demonstrably fictitious? (If you don't think the effect of the religion is entirely good or benign, you may react differently.)

When a sane, educated person seems steadfastly irrational on a particular issue, it is probably not that he is irrational, but is rationally pursuing a different end from the one you had assumed. Driving east is only irrational if you actually believed him when he said that he wanted to go west. Mysticism is a reliable way to feel consolation and hope and community; belief in alien abduction is a reliable way to relieve the boredom of our lives; going to a fortune-teller is a reliable way to feel the satisfaction of having an answer. If we recognize what it is people are really looking for, we will understand that most people are acting rationally.

Imagine the daughter who comes home from college for the holiday. She is an atheist, whereas her parents are literal in their religion. They have had many discussions about their differences, some painful, but they have discovered that they agree more than they disagree, especially on the important points. Similar to many atheists, she believes in the highest standards of ethics, and she finds the world astonishingly beautiful and the work of an incomprehensible natural system. She is enormously grateful for her existence and all its blessings, even if she is comfortable with not anthropomorphizing the object of her gratitude, which is the natural world. She sees death as dissolution only of her consciousness and her body, but that its component parts will last forever as atoms in the soil and water and air, as they have existed since those atoms were fused in stellar reactions. Similar to many religious fundamentalists, her parents believe that moral behavior, the origin of the universe, and the afterlife are all the product and domain of a conscious and mysterious God.

Despite the past disagreements, the daughter, whom we shall call Michelle, and her parents have realized that they are saying the same thing, but in different languages. They acknowledge that they explicitly disagree on some particulars, but they accept the general principle that if I have the freedom to choose my beliefs, I have to allow the other person the same freedom. So, on those particulars, they have agreed to disagree and no longer attack one another. But on the general concepts, they agree and are not distracted by the semantics:

"Michelle, thank you for coming to church with us today. That means a lot to us."

"You're welcome, Dad. I enjoyed it—the building, the community, the music. It is such a celebration of life. There is a powerful sense of unspeakable awe and gratitude that I find deeply moving. Everyone needs to take time to think and to connect to the bigger things."

"Yes. You know, we don't have exhilarating mountaintop perches in Commerce City, but the church is almost as good a place to feel that direct connection."

"Dad, I liked what the minister said about keeping the Bible in your back pocket at work, so that when someone asks you to do something, and you aren't quite sure it's in agreement with what you think is the right thing to do, you can sneak away, think about it, and make sure. I've been trying to do that lately, but life happens so fast sometimes. I need to be better at that."

"You don't have a Bible."

"True. I don't have pockets either."

Her dad laughed. "The part I liked was the kind words for Mrs. Tamaso. She so loved this church and its community center. It's a great feeling, knowing that while she's gone back to where she came from her soul is still with us, all around us. And when I work at the community center, I know she is there."

"I agree. I learned a lot from her. And the community center is a great place to be reminded of the things she wanted to accomplish, large and small. If we continue to work on those things, in her memory, it's as if she's still doing it."

"And you don't believe in life after death?" he said with a smile.

Few people have the strength to make truth their priority. Most people are far too busy with the urgencies of life, seeking simple survival, occasional pleasures, and a bit of actual happiness. That they would also be truthful in their daily practicalities is asking enough; but to be truthful in their philosophies and perceptions of their place in the world may simply be too much. Few have the luxury of so much strength. It would be a very hard person who realistically expected this from everyone all the time. Whether we expect such from ourselves and from those who represent us in a democracy may be another matter.

Consider the 6 billion people on the planet now. The range of their income, education, and worldview is enormously broad, but the sad fact is that the man at the statistical median would be of shockingly modest means. Let us consider that man as our speaker:

"You are right, Steve, truth is very important. Critical even. I do not contest that. But for the love of God, have some compassion. I am tired. I am worried. I am afraid. I have human needs. Right now, I just want to feel better. I have time and energy for nonessential thoughts only if they somehow make me feel better; it's all the better if the more I believe in them, the better I feel. Don't take that away from me. There are lots of appealing ideas that make me feel a little better. You know what I mean—the same way other little things can make you feel surprisingly better: a song on the radio, a quarter on the sidewalk, a flower on the table, a pretty girl passing by. I feel the same way when I think about seeing my parents again in heaven; about the reception I will get when I arrive there; the satisfaction I will feel when those corrupt officials meet their Maker; the praise I will get for having resisted so many temptations. I enjoy the excitement of imagining the aliens that are watching us; the healing power of the crystals in my room; the telepathic connection I have to my sister. Even if I do suspect that my notions do not hold up well to rational scrutiny, when I sense this, I just look the other way. Is that so bad? Who have I hurt? I love these ideas. I need them. They are simple pleasures in a life that hasn't enough. Why would you take that away from me?"

I'm not sure I would. It depends on whether anyone gets hurt when we think that way. If the person wants to fly passenger airplanes, build bridges, do surgery, or make laws based on their cherished irrationalities, then that may be a very big problem. If they wish to exercise strict rationality in all those realms but maintain cherished irrationalities in their private life, with no adverse consequences for others, and can keep the irrational mode of thinking from spilling over into the rational issues, then there may be no problem.

Society needs people to be more rational, but we may also need to cut people a little slack because of their humanity. Let's look more closely at some of those reasons for private, irrational beliefs.

Chapter Six
The Human Factor

Life is hard. Nature follows physical laws regardless of our preferences. It is in human frailty to not always be up for the task of facing evidence and reality. There are many worthy reasons why we can understand and tolerate irrational ideas and the people who hold them. In short, the many pitfalls and deceits, by self and others, enumerated in this book as a whole conspire to make it very difficult for most persons to believe only what valid evidence supports. Foremost, it is more common in human nature to believe whatever we must to get along in life, to believe what simply feels good, to believe whatever makes us feel righteous and noble, and to believe what sets us above other people and other animals, rather than to try to extract oneself from family imprinting and cultural saturation, set aside self-interest, laboriously seek the truth, and hold it tentatively until we find another discouragingly valid and unpleasant fact and again have to change our mind. Thus, we understand and, depending on what is at stake, tolerate these irrationalities in most persons primarily out of compassion for the limits of humanity.[1]

The Ugly Facts of Neurobiology

Recall our analogy between the nourishment or malnourishment of our bodies with food versus the nourishment of our minds with ideas. An insight by the Earl of Shaftsbury, translated into contemporary English, adds another dimension to the analogy. People process ideas much as they process food: they naturally take in ideas, digest them, use them, and eventually release them in an entirely new form, with the eventual output depending greatly on the ideas already stewing and percolating in that person. Thus, the ultimate discharge will vary considerably depending on one's health and previous ingestions. Yet however repellent those discharges may be, they are all the more unpleasant when suppressed, the delay risking serious personal and social repercussions.

It was heretofore the wisdom of some wise nations to let people be fools as much as they pleased, and never to punish seriously what deserved only to be laughed at, and was, after all, best cured by that innocent remedy. There are certain humors of mankind which of necessity must have vent. The human mind and body are both of them naturally subject to commotions: and as there are strange ferments in the blood, which in many bodies occasions an extraordinary discharge; so in reason too, there are heterogeneous particles which must be thrown off by fermentation. Should physicians endeavor to absolutely allay those ferments of the body, and strike in the humors which discover themselves in such eruptions, they might, instead of making a cure, bid fair perhaps to raise a plague. ... They are certainly as ill physicians in the body-politic who would needs be tampering with these mental eruptions. ...

—Earl of Shaftsbury (1671–1713),
Characteristics of Men, Manners,
Opinions, Times, etc.

The Appeal of Familiarity

The ideas of our youth we carry into adulthood unquestioningly. When we encounter a different set of myths and superstitions equally and transparently absurd, we quickly react with condescension, amusement, ridicule, scorn, and detailed refutation. Yet when we reflect on our own bizarre myths, we see nothing that is not obviously true, reasonable, and noble.

"I already have an answer that works for me, I don't need a new one." This is somehow reminiscent of Newton's Laws in mechanics. People's minds have *inertia*—they will sit in one place indefinitely until moved by some force. Absurd ideas of youth will stay put unless actively dislodged. Environments that are intellectually barren or that screen out new ideas never provide those necessary forces. This is why books, travel, and diverse people push minds further and why the lack of them lets a mind settle deeper into developmental arrest.

Also, Newton's Third Law states that any force applied to an object will experience an automatic force against it, generated by the outside force itself. People are much the same way, reflexively resisting an unsolicited force. We abhor the possibility, and especially the unsolicited criticism, that we have been wrong.

Furthermore, people also have *static friction*—it will take a relatively larger force to initially budge the mind from its resting place than it will to keep it moving; it is the first lurch out of a settled opinion that is the most difficult. (Neuroscientists with a mechanist bent take this notion further than you might think!)

People become most familiar with those ideas in which they were raised and are culturally immersed. There are reasons that you are not much different from the people around you—your language, clothes, lifestyle, and probably your beliefs. We like to think that we are independent—rebels even. We can always find surprising cases, but let's talk probabilities: What percentage of suburban Americans choose the Shinto religion? What percentage of ranchers are vegetarians? What percentage of Scandinavians are fervent Muslims? In both families and cultures, generally, the apple doesn't fall far from the tree. Our beliefs are far more an accident of geography of birth or the biases of our parents than they are our independent choice. In adulthood, when we confirm our beliefs as our own, we usually confirm exactly as our culture around us determined that we would. And when you change your mind, it is usually because the people, teachers, and books around you have changed. So select carefully the people, teachers, and books in your life, as well as the background babble from the television, stereo, or computer, and make sure that every decision is truly yours.

Through cultural saturation with predominant ideas, other ideas are at great disadvantage in competition for your mind. Consensus breeds unquestioning satisfaction and stifling expectations of conformity. New ideas, and anyone who seems to like them, may be treated with disappointment, contempt, or cold alienation. It is a rare person who will have significant and frequent exposure to the other ideas, feel supported in pursuing them, and develop those ideas fully. Few will swim against the current. Far more commonly, the dominant culture makes the person, and the person will resist change vigorously. The person who is certain of his opinions in this culture would be just as compliant and just as certain in any other culture in which he might have been born.

Plato observed the effect of cultural saturation about 2,500 years ago as it pertained to notions of history and morality. (The following quote refers, for example, to the story of the founding of the city of Thebes, where, the faithful maintain, Cadmus slew a dragon and planted its teeth, which then grew forth as the first noble families that founded the new city.) In Dodds's words:

> In [Plato's] Laws, at any rate, the virtue of common man is evidently not based on knowledge or even true opinion as such, *but on a process of conditioning or habituation by which he is induced to accept and act on certain "salutary" beliefs.*[*]

After all, says Plato, this isn't too difficult: people who can believe in Cadmus and the dragon's teeth will believe anything.[2] Far from supposing, as his master had done, that "the unexamined life is no life for a human being," Plato now appears to hold that the majority of human beings can be kept in tolerable moral health only by a carefully chosen diet of ... edifying myths and bracing ethical slogans.

*It was I who added emphasis.

People are most familiar and agreeable with the ideas that they have authored themselves—"ownership bias." In fact, a reliable way to guarantee opposition to your idea is to exclude others from participating in its formulation—this is human nature's fear and distrust of the unfamiliar.

> "Men often oppose a thing merely because they had no agency in planning it, or because it may have been planned by those whom they dislike.
>
> —Alexander Hamilton
> The Federalist Papers, No. 70

Many teachers themselves still believe that familiarity ingrained through repetition and saturation are the keys to "education"; one of their (our?) first rules of classroom presentation is "Tell them what you are going to say, say it, then tell them what you said." Others might worry that this would do more to promote memorization and blind consensus than education. The point is that through a variety of individual and cultural mechanisms, ideas become familiar and exclusive, and it is human nature to confuse familiarity and exclusivity with validity.

Last, just as we underestimate that which is unfamiliar, human nature causes us to overestimate that which is familiar. Suppose I've read ten good books on a particular issue and you've read one that you found exceptionally good. If I have never even heard of your book, you may leap to the conclusion that I am uninformed or obviously biased. You cannot imagine that I could be well informed without the source that is so central to you, because if that one source were removed from your life, there would indeed be a huge hole until you filled it with other sources. (See Fear of the Mythical Void on page 171.)

Fear of Change

We may resist new ideas in part out of fear of a change in "who I am." We know that the change will be quickly noticed by friends and family and that they may well have a strong reaction; we fear the possible commotion, ridicule, and rejection. Besides, we are quite accustomed ourselves to our own ideas; we ourselves do not want to leave behind the person we have always known and become a person who is different and perhaps a little more similar to some people we had long criticized.

It is painful for most people to discover that the weight of evidence falls fairly convincingly in favor of a conclusion that is not the conclusion they had supported until then. No one likes to be wrong, or to discover that the people they love, respect, and wish to please are wrong, or to have to admit that they themselves had previously operated with insufficient understanding, or that they have learned something from their opponents and detractors. Humble pie is tough chewing and goes down slowly. It is a human tendency to form camps or tribes around common beliefs—customs, religious dogma, political opinion—so one who changes his belief in favor of the other can be branded a traitor; he may feel as though he is a traitor himself, with no one accusing him but the voices in his memories: "I can just imagine what my mother would say. ... "

To the contrary, many feel that it is not treasonous to change one's mind, but that it is one of the most courageous and noble acts that a person can make. It depends on the reasons for the change.

> Foolish consistency is the hobgoblin of little minds.
> —Ralph Waldo Emerson, "Self Reliance"

It may be all the more scary to realize that one's position is untenable and there is no obvious alternative.

Fear of the Mythical Void

The loss of your answer does not mean there is no answer; indeed, it may be a necessary step to discovering a better answer. Dismissal of your solution does not mean there is no solution at all. Having never studied the alternatives, we suppose *they* are weak and ephemeral when in fact only our *understanding of them* is weak and ephemeral: "Doctor, if you tell me that I cannot eat pizza, hamburgers, chips, or French fries, I guess you want me not to eat at all"; "You don't watch TV—what do you do all day, sit and stare at a blank wall?"; "If we do not have *this* law, then there will be lawlessness"; "If they do not have *this* moral code, then surely, their foot shall slide into immorality"; or "If we do not have this book to tell us right from wrong, then we'll have no compass at all." The void is a myth because there are always alternatives, and we have the responsibility to investigate and assess them.

Moving from stable, unearned certainty to honest and demanding uncertainty is difficult and unsettling, especially in a culture that views uncertainty as a moral or intellectual failure rather than as a legitimate and wise stance given valid but conflicting considerations. Consider, for example, the popular aphorism "If you don't stand for something, you'll fall for anything." Catchy, yes. Cute, yes. But dead wrong. For starters, it is a flawed syllogism: to believe that you will be easily duped means the speaker makes the unstated premise that you are incapable of making careful and discriminating judgments. It advises you to be dishonest with yourself and others—that you choose to "stand for something" now, even if all available options are unsatisfactory to you; he apparently wants you to feign certainty. Perhaps I am too harsh—would he accept it if you vigorously "stood for" withholding judgment for a reasonable time while you studied the issue further? Tell him that you stand for proactive collection of all sensible considerations, for patient assessment, for reason, and for the right to change one's mind unapologetically as new information arises.

Fear of Freedom and Responsibility

The burden is too much—the responsibility to be curious on all matters, to sweep aside all prejudices and baggage of childhood, to seek out advanced education inside and outside of school, to cultivate a circle of thoughtful peers and challenging mentors who will serve as a guide to the valuable insights, to learn the specific skills of sorting through library databases and an ocean of publications, honing the skills of critical analysis, statistics, science, and rhetoric so one may be able to weigh the degree of validity of each piece of evidence. Many people do not want the freedom to think for themselves or the responsibility that comes with it.

It has happened before in history that a whole society evolved a respect for this freedom and responsibility, then they fled from it. The rationality of the ancient Greeks was abandoned by the Greeks themselves, as they chose instead the relief offered by an authoritarian social system that agreed to take over for them the thinking.

Behind such immediate causes [for the Greek public of 200 B.C.E to reject rationalism in favor of mysticism and unfounded certainties] we perhaps suspect something deeper and less conscious: for a century or more the individual had been face to face with his own intellectual freedom, and now he turned tail and bolted from the horrid prospect—better the rigid determinism of their

astrology than the terrifying burden of daily responsibility.
—E. R. Dodds, The Greeks and the Irrational

Lying

This is a natural human tendency that requires no explanation. However, it deserves to be noted that in lying, we see reason's evil twin: rationalization. Here, reason has been co-opted by some goal other than truth, and for the sake of some gain that we may hide even from ourselves, we dress up our rationalizations as reason and parade them exhaustively to assure everyone, including ourselves, that certainly this is reason at work. Leo Tolstoy's own personal conflicts about a luxurious life built, in part, upon the toil of impoverished serfs poignantly reflect how we may lie to ourselves:

> I sit on a man's back
> Choking him and making him carry me
> and yet assure myself and
> > others that I am very very
> > sorry for him
> and wish to lighten his load
> by all possible means—except
> by getting off his back.

Weasel Words

When grown-ups lie: "I did not have sexual relations with that woman, … "[3]; "I did not declare war without authorization from Congress—this is a police action"; "I did not steal that money from the shareholders; they gave it to me to do what I thought was best"; "When I said yesterday that I didn't do that, I meant I didn't do that—*yesterday*"; or "But the police said I was seen beating up some guy in a blue shirt, and I didn't—it was more purplish."

A famous example of weasel words is seen in Pascal's *Wager*. Blaise Pascal was a seventeenth-century French philosopher and mathematician who lived in a time in which people were pressed, despite their personal judgment, to state their conformity to the dogma of the times, which included a belief in the doctrine of God, judgment, heaven, and hell. Pascal reasoned that if he chose to state such a belief in God and was wrong, then he had lost little, because after death there was, in fact, only oblivion. On the other hand, if he chose to reject such a cosmic system and was wrong, he would pay the price with an eternity of agony in hell. Thus, he chose to "believe" and recommended that the rest of us tell authorities and God that we believe. It is easier and safer to believe.

But critics of Pascal's *Wager* argue that oaths made under coercion, while possibly satisfactory to earthly authorities, would probably not impress loftier ones, and that uttering the words as if they were a magical incantation that would pop open the pearly gates regardless of sincerity would probably not work with an omniscient gatekeeper.

A sweet religion, indeed, that obliges men to dissemble and tell lies, both to God and man, for the salvation of their souls!
—*John Locke, Letters Concerning Toleration, 1689–1693*[4]

Even more impressive is that many persons are capable of making Pascal's Wager without even admitting to themselves the insincerity of what they do. They do not genuinely believe, they are full of questions and uncertainties, but they feign believing, even to themselves, going through the motions, uttering the words, and swearing that they believe, even to the point that they may convince themselves in moments of emotional certainty that they do believe. When skeptical questions occur to them, they avert their eyes, close their ears, distract themselves, and do anything else it takes not to follow the thought, lest they find it reasonable. They are capable of choosing to believe what is in their interest to believe, rather than what the evidence indicates is true. And they are capable of believing this is sensible.

Some make the world think that they believe what they do not; others, in greater number, make themselves think that they believe what they do not, not knowing what belief is.[5]
—*Michele Eyquem de Montaigne (1533–1592)*

Suppressed Evidence and Half-Truths

Sceptics of the theory of evolution frequently make the pseudoscientific argument that "The law of entropy proves that the natural tendency is toward disorder, not order, thus evolution is impossible." But they intentionally leave out the law's phrase " ... unless energy is put into the system," such as the sun and the competition to survive. "Did you bring me all the change after you bought the milk?", "Well ... yes." "So where did the candy bar come from and why was the milk so expensive?"

The Appeal of Money, Fame, or Prestige

Magicians, faith healers, priests, previous-life channelers, Dial-A-Psychic ... all of these are big businesses. Some people and institutions sell breast implants because some women feel better to have larger breasts; other people and institutions sell the reassurance that humanity was created by omniscience and that omniscience loves you and worries for your welfare, because some people feel better believing that. The market is huge; a great deal of money and status is at stake. These services exist in part because there is a demand for them.

People are holding out lots of money asking for answers, awe, consolation, hope, and community, so all types of people step forward to sell those things to them. The products supplied, generally speaking, come in as many flavors and styles as do the demands, and people are free to use their own money, time, and identity to invest in the product that is most satisfying to them. Crop circles, healing magnets, Roman Catholicism, and evangelism are tailored to meet particular tastes. The seller has the earnings and prestige of providing something enormously valuable; the buyer has the comfort of being elevated to a higher level than the reality as she originally found it. We can do better than that.

Academic research can also dish out whatever is called for. Students, faculty, and many professionals may conduct "research" with entirely predetermined conclusions. They collect only the evidence that supports their conclusion and that refutes the alternatives. They often encounter all the opposite pieces of evidence, but ignore them. Why? All too often they are writing the paper for a paying client who expects a very particular conclusion. In other words, they defend themselves with "I was just doing my job." Supply and demand. We can do better than that also.

Wherever people in need are willing to offer money for a solution, someone else will be willing to meet that need, whether the "solution" is valid or not, whether the providers of the services believe it themselves or not. Again, truth takes a backseat; truth may not even be on the bus.

Consider also the opposite appeal. ...

The Appeal of Appearing Humble

We see here that any respected trait can be a laurel prized more highly than truth. In this case, a person may proudly state that they choose not to question the apparent irrationalities inherent to their notions because it would be arrogant to put one's reason up against that of the great party fathers (or any other revered source); better to stay quiet and enjoy being praised for boundless humility.

Pride, perceiving humility to be honorable, often borrows her cloak.
—*Thomas Fuller,* Gnomologia, *1732,*
as quoted in Graham Hancock's Lords of Poverty

Then again, maybe it's not really so humble to believe that one has tapped the knowledge of omniscience or is the deputy of omnipotence.

The Appeal of the Fantastic

Mystical, fearful, supernatural, and exciting ideas are pleasant. They are popular precisely because they are extraordinary and offer relief from mundane life. They are often far more interesting than the monotony and predictability of natural law and the real world that comes with it. That's one of the reasons why we like movies so much.

- "What do you think lies beyond the farthest point in the ocean?" Monsters! Mermaids!
- "What was that bump in the night?" Ghosts!
- "A weird light in the sky! It must be an alien spaceship!"
- "I had an amazing dream last night. I think it's a message. ... "

The appeal of the fantastic is also one of the reasons we create and adore celebrities. Anyone who has tried to play basketball will watch Michael Jordan and think that he somehow defies natural law. Our twelve-year-olds utter a very human response when they call him a basketball god. No one, it seems, should be able to hang in the air that long or make that last-second, game-winning three-pointer despite smothering coverage. He seems to deliver miracles at will. When watching him play, we have the thrill of suspension of natural law. Because his play is *fantastic*, we are *fanatics*, or *fans*, of him. And he, in turn, is a celebrity, a word that is derived from the Latin *celebrare*, "to solemnize," "to glorify." This same Latin word gives us *celebrant*, a participant in a religious rite. And, as our ancient ancestors continually associated the heavens with their gods, we still call our celebrities "stars." These people temporarily transport us from our lives that are otherwise ordinary and hopelessly bound by natural law.

The appeal of the fantastic is also seen in the ancient worship of pagan gods. In the fifth century B.C.E., Greek devotion to the god Dionysus, for example, was based on many aspects of escape from the real world. A follower lived with the burden of original sin due to ancestors who had offended the gods, but this could be relieved through devotion and through the "communion" of partaking of the body and blood of Dionysus in ritual-istic fashion. The follower lived in fear of an eternal afterlife in hell unless he appealed to Dionysus with penances before death or his friends performed them for him after his death. Dionysus was a person who had personally defeated natural law and who offered relief from natural law through the performance of miracles—in one story of his life on Earth, for example, he had only water to drink, but was able to turn it into wine.

Devotion to Dionysus was based primarily, according to Dodds, on the relief from the real world that he offered: " ... not only because life in that age was often a thing to escape from, but, more specifically, because the indi-vidual, as the modern world knows him, began in that age to emerge for the first time from the old solidarity of the family, and found the unfamiliar burden of individual responsibility hard to bear. Dionysus could lift it from him." As Will Durant observes, "For Dionysus could make a vine grow out of a ship's plank, and, in general, enable his votaries to see the world as the world's not."

There are plenty of examples of such escapism in today's world: TV is one.

The Appeal of Basic Necessities

The Romans also realized that the masses' need for entertainment was pro-found, as real as the need for food, and the consequence of not providing both could be serious. While we know better than to take a satirist as an his-torian, consider for yourself if there is truth in this insight:

> Duyas tantum res anxius optat—
> Panem et circenses.
> *The people long eagerly for two things—*
> *Bread and circuses.*
>
> —Decimus Juvenal (circa A.D. 60–140)

Today, the human need for these is undiminished and may often be strong enough to overwhelm rationality. If a person finds certain beliefs profitable or escapist, he may have no trouble believing.

The Appeal of a Higher Meaning, Purpose, or Comfort

Beliefs might offer relief from a frightening view of the universe. Some people have observed the natural world's innate creativity, its faceless mechanisms, and its apparent indifference to humanity, and have been deeply troubled. Perhaps what the humanists say is true—that humanity exists only by an accident of physical events, that there is no special honor intrinsic to humanity, that there is no soul, that bad deeds may go unpunished, good deeds unrewarded, and that, for the individual, after death there is only unsensing, undreaming oblivion.

> A man said to the universe:
> "Sir, I exist!"
> "However," replied the universe,
> "The fact has not created in me
> A sense of obligation."
>
> —Stephen Crane (1871–1900), poet

In such a world, what is the meaning of life? One would have to ponder and seek and struggle with uncertainty and answer for oneself and perhaps be wrong and have to change one's mind—perhaps many times—and risk being uncertain, even in old age, with one's regrets and one's fears and one's deathbed.

For some, such an existence would be intolerable, the idea itself intolerable. They can be desperate to substitute a more pleasant view, regardless of its foundation in science or psychology. They want a meaning and a purpose that make them feel good; forgiveness for their cruelties when they can't bear to, or are unable to, get it from the person they had hurt; certainty that they are loved by someone; certainty that their birth, life, and death will be acknowledged and regarded as meaningful. And they want it now, ready-made, rather than to have the responsibility and burden of searching forever for meaning, purpose, and comfort within that awful context. Dodds again: "The Greeks believed in their Oracle not because they were superstitious fools, but because they could not do without believing in it."

And who would blame them? For God's sake, would you deny it to them?

Of course, human nature knows that this is an opportunity: as always, people will step up to supply that demand. This is satisfying to both the giver and the receiver.

> Credo Consolans. More than any other, the reason people believe weird things is because they want to. It feels good. It is comforting. It is consoling ... to the frequently asked question "What is your position on life after death?" my standard response is "I'm for it, of course." It is a very human response to believe things that make us feel better.
>
> —Michael Shermer,
> Why People Believe Weird Things

The Appeal of Social Order

We often hear the alarmist claim that a proposed change would "destabilize society" or "lead to social chaos." As often as this is claimed, and as rarely as it actually occurs, it appears to be an unlikely result of most new ideas. Perhaps we perceive unsettling change to a comfortable status quo as "chaos" when, in fact, it is simply the normal stress of change. Yet there are also plenty of examples of social chaos in human history, so we do well to consider the risk. We also do well to consider the reward of a bit of temporary instability, for if we have an exaggerated commitment to social stability or an exaggerated fear of instability, we may miss opportunities for vital change. Stability and change must be balanced optimally.

Although social stability has long been a goal of leaders, many of the great social advances came after, and perhaps because of, a time of frightening and terrible social instability—the Magna Carta, the Reformation, the American Revolution, the American Civil War, labor unrest and strikes, the collapse of colonialism, the civil rights movement, to name a few. And, even without violent upheaval, some fear the "chaos" of differing opinions spoken and lived, and of change occurring in self, family, and societies, and so would rather have familiar stability. Yet progress requires, it seems, periods of instability—so long as the risks of the instability justify the potential gains in truth and justice.

Rationality at what price to social order? Social order at what price to rationality? Are the two really at odds? For at least 2,500 years, intellectuals and philosophers have reluctantly accepted the irrationality of the masses because they have concluded with sadness that it would be individual cruelty and social disaster to take it away from them. (See the Dodds's quote about Plato on pages 172–173.) If an individual changing can be so convulsive, they reasoned, just imagine what might happen to society as a whole. If irrationality confers some consolation, would rationality threaten society with unmitigated despair? If irrationality confers some meaning, would rationality eviscerate us? If irrationality gives us reason to improve our individual moral behaviors, would rationality unleash our dangerous impulses?

The ancient intellectuals came to suspect that civil society might do worse than with the irrational mysticism of the time. Perhaps there is rationality in letting the masses be irrational.

*Since the masses of the people are inconstant, full of unruly desires, passion-
ate, and reckless of consequences, they must be filled with fears to keep them
in order. The ancients did well, therefore, to invent gods, and the belief in
punishment after death.*

—*Polybius, 123* B.C.E.*, Roman historian,
as quoted by Carl Sagan in* Demon-Haunted World

Centuries later, when Europeans could speak less freely, Voltaire, in a
letter to Frederick the Great, prince of Prussia, made the same point, more
safely: "If God did not exist, it would be necessary to invent him."

Many people, at risk of being called naïve, do not agree that people
must be allowed their religion, however irrational, for the sake of social stability. However, they would not try to forcibly suppress a person's religion,
which strikes them as cruel and as a violation of civil rights, but they would
hold up rationality and hope the others would see it, know it, and try it.
Letting go of irrationality may well leave a sense of void, but it is that sense
of void, the "I don't know," that has always been a necessary step before further
investigation is begun. That further investigation may lead to a worldview
that is both rational and has the benefits of consolation, meaning, and moral
restraint. An individual changing need not be convulsive; neither a society.
Irrationality can fall away slowly and gently, just as the elk's winter fur coat
falls away—when the time is right, when it's not needed anymore. We may
have to let it happen gradually.

*If they're beautiful I don't mind if they're not true. It's asking a great deal that
things should appeal to your reason as well as to your sense of the aesthetic. ...
Perhaps religion is the best school of morality. It is like one of those drugs you
gentlemen use in medicine which carries another in solution: it is of no effi-
cacy in itself, but enables the other to be absorbed. You take your morality
because it is combined with religion; you lose the religion and the morality
stays behind. A man is more likely to be a good man if he has learned goodness through the love of God than through the perusal of Herbert Spencer.*

—*W. Somerset Maugham,* Of Human Bondage

*The uncertainty and vanity of the mind are such that it must always enter-
tain an opinion: it is a child that must be presented with a toy so that we can
take away a dangerous weapon; he will lay aside the toy himself when the age
of reason is attained.*

—*Jean Le Rond d'Alembert (1717–1783), introduction to* Encyclopedie

And yet, often the adults never do lay aside the toy.

To walk on the water in fancy is much less wearying than to go forward in earnest on the roads of the earth.

—*Simone de Beauvoir,* The Second Sex

Why take the potential risks that may come with the loss of myth? Because even limited efforts to replace mysticism with rationality have brought forth incalculable improvements in medicine, law, and numerous other fields. Our future with rationality is terrifically promising. We can't see the future to know what rationality would do for humanity on a large scale, but we have never been able to see the future with any great endeavor. If we had allowed ourselves to be content with hunting and gathering, we would have never advanced to domestication and agriculture. If we had allowed ourselves to be content with snake oil, we would never have discovered antibiotics. If we allow ourselves to be content with abject mysticism and withdraw from the freedom to think rationally, we will never discover the evidence-based approach that can unshackle our minds and free us to fulfill our human potential.

We have seen now the many ways in which we can fool ourselves and others. If we are instead to insist on hard evidence and if we are to be competent to measure fine distinctions of valid data, then we must have a way to quantify the data, and especially to quantify our uncertainty. This is exactly what the science of statistics does. Despite the pervasive fear of mathematics in contemporary culture, statistics is an approachable, comprehensible, and indispensable tool for a rational thinker. In the next chapter, we will look at some of the major concepts in the field of statistics, in hopes of inspiring the reader to continue on to a truly thorough training with the incisive skill.

Chapter Seven

The Other Literacy: The First Ten Lessons of Statistics

People who are searching for valid answers to their questions will often encounter statistical studies. Some people are reasonably comfortable in the green fields of *words*, fairly well trained and experienced in finding their way through prose, even when it's tricky. But *math*? Some of those researchers will come to a stop. Staring into what seems to be a deep forest of tangles and uncertainty, they chicken out, move around the statistics, and head back into the prose with which they are comfortable, hoping that realm alone will provide sufficient evidence.

It is not okay to chicken out. You need what is in those mathematical studies. "But no one understands higher math—no one would blame me if I didn't put myself out either," they may say. Sadly, there is some truth to this. On the one hand, most of us live in a culture in which individuals are expected to be literate; it is not socially acceptable to say, "I'm not good at reading. I don't do reading." No one considers such a statement acceptable, amusing, or endearing. But, on the other hand, we are not surprised or offended to hear someone say, "I'm not good at math. I don't do math." That's not okay: those who want to be able to distinguish good ideas from bad and who want to reason well must be numerate—literate with numbers. As being literate means more than comprehending "See Spot run," being numerate means more than comprehending long division. It means comprehending statistics.

This book has thus far emphasized the role of uncertainty in decision making. Statistics can be described as a method of quantifying the degree of uncertainty. (At least, I'm 95 percent sure that's what statistics is, plus or minus a few percentage points.)

Statistics get a bad rap. Some people are so used to hearing statistical fragments thrown around that they have given up on believing any statistical information at all. When they hear statistics such as "Four out of five doctors surveyed said, ... " they rightfully wonder if only five doctors total were surveyed; even if 1,000 doctors were surveyed, they wonder if there was bias when those doctors were selected from the larger pool of doctors available. Good questions. Having expressed healthy skepticism, they retreat into cynicism—they believe *no* statistics and may abandon an attempt to understand statistics generally. This is a great loss, because many of life's most interesting questions can and should be informed by studies whose validity is determined by and scrutinized through statistics. Misleading statistics will be with us always, just as misleading prose and misleading authorities will be. But we must not give up on statistics any more than we give up on reading or on human institutions. Instead, we must learn to reason through them, to separate the good from the bad, and to weigh the degree of value, the strengths, and the weaknesses of each.

The numerate, then, are those who find their trail of evidence leading into the different terrain of mathematics and are undaunted. They are well equipped to charge right into the statistical studies, if that is where their evidence is. They will know how to read the studies, they know what to look for, they know how to weigh the validity of the conclusions, and they know whether the little statistical fragments that are thrown about from the study are valid or not. They know that no study is perfect, but they know how to make appropriately tentative conclusions from the best data available.

This chapter surveys the essential landmarks in basic statistics and highlights what I consider to be the top ten lessons. If you can follow me through these ten lessons, then you should have no fear of mastering statistics through comprehensive study. This chapter is no substitute for a course or book on statistics, even an introductory one, of which there are many.

With full admission that I am oversimplifying what is not simple, I submit this cursory orientation to statistics in hopes of inspiring the reader to study it in detail and fill in my many sins of omission.

Organizing Raw Data

Statistics starts with raw data. For example, if we asked a group of ten people the number of pens they each had in their backpack, we may record our data this way:

Observation (person): 1 2 3 4 5 6 7 8 9 10
Answer (datum): 22 9 12 32 35 42 8 12 98 39

As data sets grow larger—such as when this same question is put to the first 200 students to walk out of the library on a Friday evening—then it may serve to represent the data set as a histogram. A histogram is a type of

graph with the possible answers running along the bottom and a dot for
each person who gave that answer stacked up on top of that answer:

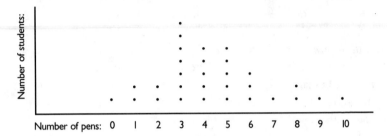

It is not hard to imagine that the shape of the histogram could be anything, depending on the data set. What would you imagine the shape to be
if 200 persons were asked:

- How many pens do you have? (walking out of nudist colony)
- What is the value of your home in Puerto Rico?
- How many wheels were on the vehicle that you took to school today?
- Which letter does your first name start with?
- What was the sum of the two dice you cast?

The first question, of course, makes the point about knowing the
potential bias of your population selected for the survey.
Consider that last question: What would be the shape of the histogram
if only ten people threw the dice once each? Thirty people? Two hundred
people? In fact, with true dice there is only one way to get 2: with a 1 and a
1. It's the same chance with getting 12. But there are six ways to get 7, and a
gradual gradation for each answer in between, so the histogram for a large
number of throws would start to take on a shape that reflects the inherent
probabilities:

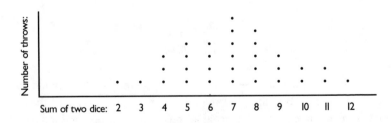

The fact that only ten or twenty throws will not produce the obvious overall pattern reveals a very important truth in statistics:

Lesson number one: If one claims that there is a pattern evident in the data, then the size of the data set must be large enough for the pattern inherent in the system to be evident. Just how many observations are necessary for your conclusions to have a sufficiently small degree of uncertainty is a manageable calculation. We shall explore that question further when we get to "sampling."

Measures of Central Tendency

One simple way to describe any data set is with its "average." The average is intended to represent a number that describes the center of the data set, but "average" can legitimately mean any of three different such "measures of central tendency": the mean, median, and mode. Why the semantics? Because the difference is, well, statistically significant.

Measures of central tendency:

Mean

Arithmetical average: add up all the data points, divide by the number of data points, and you arrive at a number that is, in one sense, "central" in the data set. In the case of the number of pens on page 183, the mean is 30.9.

Median

The central datum: if the data are arranged in order of increasing frequency, then what value is in the center physically? (Imagine that with the list below you were to throw out the first datum on the extreme left, then the last datum on the extreme right. Repeat. Repeat until there is only a single datum remaining. That is your median.)

8 9 12 12 22 32 35 39 42 98

In this case, there is an even number of values, so we find in the middle, 22 33. We take the mean of those two: the median is 27.5.

Mode

This is the datum that occurs most frequently, in this case, 12. Modality is useful when the value that occurs most often is a more meaningful descriptor of the data set than the other types of "average."

If the data are distributed in a symmetrical fashion, the mean, median, and mode will all be the same. If the curve is skewed one way or the other, these measures are likely to be different from one another.

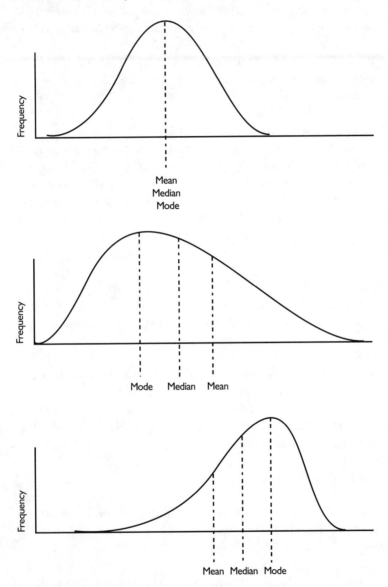

Some data sets are clustered around the mean, as in the example above with the dice. But if you had no familiarity with dice, and you had asked what the average was when two dice are thrown, would "7" be adequately descriptive? Probably not, because such a simple answer says nothing of the wide spread and the interesting distribution of actual answers.

Consider some other examples as well.

Different countries spend different amounts on health care, per capita. Where does your country stand relative to others? The variation between countries is quite wide and not at all a symmetrical distribution, so the arithmetical average may not be representative. In this case, comparing your country's figure to the *median* may be more illuminating.

In some countries, it is possible that there may be a large number of humble homes, a small number of enormous homes, and very little in between. Would the "average" price be representative? This is a case in which the *mode* may be more representative, if any one number is useful at all.

What would be the average number of wheels on the vehicle taken to school? Given that a good many cars and bicycles would likely be involved, one would not be surprised if the "average" were close to three! Is that a sufficient reflection of the type of vehicle taken to school by these students? This is a case in which *modality* is evident in the histogram.

As for the question concerning the first letter of the names, how does one "average" nonquantified answers?

Lesson number two: When you hear the word "average," ask if the person refers to the mean, median, or mode. Ask also for more information about the range of values and their distribution in that range. Better yet, don't be satisfied with an "average"; ask for the original data with the histogram.

Beyond the Average: Measures of Spread and Distribution

When you look at a histogram, what you are likely to notice is how widely the data are spread. The measure of this is the standard deviation. Don't panic. Despite the statisticlike term, standard deviation is an easy concept: it is a number that represents, roughly, the average difference between any *particular* data point and the *average* of all the data points.[1] Tall, skinny, bell-curve histograms are made of dots that are tightly clustered around the mean, so the average distance from any one data point to the average of all the data points is small. Short, broad bell curves have large standard deviations.

Lesson number three: If someone tells you the number of data points, the mean, and the standard deviation, you can have a reasonable idea of the data set, even if you are talking about hundreds or thousands of observations!

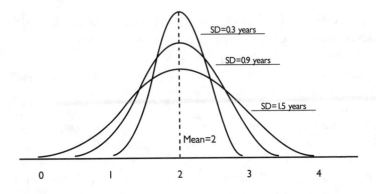

Standard Deviation, SD, of three community colleges and the typical number of years before students transfered to a four-year college. Note that each is a normal probability distribution with equal means but different standard deviations (SD).

To describe the location of a data point relative to the mean, we describe it in terms of how many standard deviations away from the mean it is. One student's SAT score may be 1.8 standard deviations above the mean, which you would interpret as being considerably better than the average score and better than the average person who did better than average (which would be only 1 standard deviation above the mean)!

In fact, according to the Empirical Rule, if the bell curve is indeed close to symmetrical,[2] then with little variation for the skinniness or the squatness of the curve, you can know that about 68 percent of the data points are within one standard deviation of the mean, and about 95 percent of them are within two standard deviations of the mean, and practically all (about 99.7 percent) lie within plus or minus three standard deviations of the mean. If a person claims that their hat size, number of romantic dates last year, or number of pieces of pizza eaten last month was more than three standard deviations above the mean, it tells you that you are indeed talking to a special person.

Lesson number four: By committing just a few pairs of numbers to memory, you can even make quantitative inferences about data that falls in a symmetric bell-curve distribution: one standard deviation, 68 percent included; two standard deviations, 95 percent included.

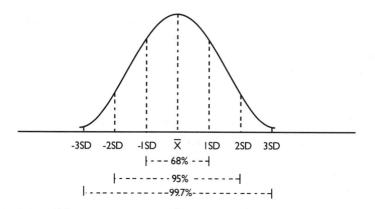

A symmetrical, bell-shaped curve showing the relationship between the standard deviation and the mean according to the Empirical Rule. (SD=standard deviation; $\overline{X}$=mean)

Surveys

Let us now look at data from a different perspective—when we are trying to ascertain the percentage of a population that has a particular characteristic. For example, suppose we are wondering what is the percent of a population that is going to vote for a particular candidate in the upcoming presidential election. The poll just told me that 46.3 percent of those surveyed intend to vote for the candidate I like. How reliable is the figure? Is it an accurate reflection of the *actual* percentage out there that intends to vote for that person? A statistician would translate that question into: I can say with ____ percent confidence (one would hope for a large number, perhaps 95 percent or greater) that the true figure lies within the range of 46.3 percent, plus or minus ____ percent (one would hope for a small number, perhaps 3 percent or less). Tacking on that bit of probabilistic hedging is the honest and quantified description of your degree of uncertainty. That buffer around the proposed answer is called the error bar, or the confidence interval.

The way in which those final numbers were derived—the sampling design of the study—is critical to the validity of the data. Following are some critical factors for validity that one must weigh.

Sample Size

The size of the sample must be sufficient for the question being asked and for the degree of confidence one desires (if attainable). Surprisingly, even if the population studied is enormous, perhaps 75 million likely voters, the size of the sample doesn't have to be. If the sample is genuinely random, and the variable is well distributed—the appearance of voters for different candidates is well mixed—as few as 1,000 persons surveyed could give a result with a fairly narrow window of error.[3] When those conditions are not reliable, then a larger sample size might compensate. In fact, quadrupling the sample size will halve the size of the interval.

However, making the sample ever larger quickly has diminishing and negligible benefits because the size of the sample can never compensate for problems that also increase along with it, such as the percent of persons who do not respond to the survey, lie to the surveyors, or change their mind before election day.

If the overall study plans to break down their results into sub-studies (such as, "What percentage of middle-class Latinos on Medicare are voting for my candidate?"), then the sample for that question alone will have to be sufficient by the rules above. Compound that with each of the other questions you want to answer, each with its sufficient sample size, and the *overall* size for the entire study begins to amplify. But that's a function of how many questions you want to answer. For each question, a well-designed study may sample only a very small percentage of the population studied.

Nonetheless, the size of the sample that is sufficient for accuracy is not hard to calculate. The energy is better invested, for both the researcher and the one who is scrutinizing the validity of the study, on considering other critical factors in the integrity of the sampling.

Lesson number five: Size matters, but not as much as you'd think. Other factors are more important, such as technique, including randomness, the confidence level, the size of the confidence interval, and the study design.

Some of those other factors critical to the integrity of the sampling are outlined below.

Randomness
Were the individuals sampled truly chosen at random? Did every subject in your population have an equal likelihood of being selected? Some specific questions:

- Were the individuals surveyed simply the ones most convenient to the surveyors? If they surveyed by phone, then they likely overrepresented persons with more than one phone line, and such a group may well answer differently overall than the general population. Conversely, phone surveys underrepresent persons with no phone or no listed phone.
- Sidewalk surveys may represent only that particular neighborhood and only

those whose lives allow them to be on the sidewalk at that moment rather than at their desks, in bed, or anywhere else.

- Did the surveyors have arrangements for persons who speak other languages? What is the magnitude of that possible error?
- Did those surveyed volunteer? Volunteers answering a survey are often very different in response to the nonresponders and hard-to-pin-down types.
- Were the responders enticed with a small bit of cash? Those responders may be quite different from those who are not effectively lured by that amount.
- Did the act of picking one individual affect the likelihood of another being picked?

Lesson number six: Always scrutinize studies for the method by which they included the subjects. Do they seem to be representative of the population that interests you? Consider some of the questions listed above. Since no study is perfect, comment on what you think the magnitude of the error is as it relates to sampling bias.

Confidence Level and Confidence Interval

In the example above, we used a 95 percent confidence level: there is a 95 percent chance that the actual percent of persons who intend to vote for my candidate is within a stated range; that stated range is the confidence interval. But we could just as well have used an 80 percent confidence level, which would have given us a commensurately smaller confidence interval to focus on. (That is, if you wish me to give a more precise answer, I can do so, but with less confidence that I am right). Conversely, we could have used a 99 percent confidence level with a larger confidence interval. Either way, we admit to our uncertainty and quantify it.

Lesson number seven: You know the authors cannot claim certainty; there must be a confidence level expressed, and a confidence interval associated with it. Find them and ask yourself if you are comfortable with them. Trust the results only so far as the study design and confidence level indicates.

Study Design

Some studies are more powerful and reliable than others. The studies are designed primarily to isolate a single characteristic—such as the intent to vote for a particular candidate or the use of a particular drug—and all other variable characteristics are eliminated, reduced, or evenly distributed across the studied populations so as to reduce the effect of these other factors on the observed difference. But, typically, the powerful and reliable studies are also much more expensive because they are laborious and prolonged. Some typical study designs are:[4]

Randomized, Double-Blind, Controlled: Randomized, controlled studies yield the strongest evidence in medicine. Randomization is the strongest safeguard against unanticipated study bias.

- Randomized: The allocation of subjects into each group is entirely random.
- Double-Blind: Two or more populations are being studied, such as a group taking a new medicine and a group taking a placebo, but both the doctors and the patients are "blind" as to who is taking what, which is known only to the researchers who put codes on the bottles. After an appropriate time of taking the medicines, the code is broken, the results for each subject are sorted into their respective populations, and the differences are compared.
- Controlled: Relevant characteristics that could affect the subjects' odds of getting the disease in question are known so that the results can be adjusted to eliminate the possibility that the observed difference would end up being a perceived result of the medicine taken. Typical confounders in medical studies are age, socioeconomic status, education level, lifestyle risks, other medicines, and other diseases.

Well-designed case-control and cohort studies:

- Observational studies: Reports of expert committees or respected authorities is considered the weakest form of evidence relative to the above types of studies, but may be better than opinions from nonexperts.

Lesson number eight: Look at the study design. Weigh the strength of the design type, as well as the strength of its execution—randomization, blindness, control.

Hypothesis Testing

"Whoda thunk? What'r the odds?" You are given a statistic that surprises you: at current rates for abortion in the United States, about one in three women will have had an abortion by age thirty-five.[5] This surprises you because you had been told that abortion was rare—you seem to recall that the "real" number is actually only 5 percent. You know that the researchers did not query every woman in the country to measure that number directly—they likely took a sample and estimated the real number. But was it a valid sample? By the critical factors discussed so far, you find nothing amiss. But how likely is it that when the "real" number is 5 percent of subjects with a history of abortion, a properly conducted sample will happen, just by chance, to have 33 percent? It could happen, just like it could happen that in an oyster bed in which the oysters with pearls are randomly and rarely distributed, a diver *could* find a pearl in four consecutive oysters—but what are the *odds* of that happening? You can calculate them. And as the odds get smaller, the suspicion that one of these two figures is wrong increases—it becomes less and less likely that in a 5 percent world, a good survey will get a 33 percent estimate. Rather, something was likely affecting one of the two studies that gave an unlikely result.

This is the essence of hypothesis testing. First, we state the null hypothesis: that there is no bias in the study and these results did indeed occur by chance alone. The reality is 5 percent, we just got unlucky with a good study that captured a sample with an unrepresentative 33 percent. The alternate hypothesis is that in the recent study, by design, women with abortion were more likely to have been represented in that particular sample than by chance alone—there was a bias at work.

The answer will read: given the assumed underlying circumstances, the odds of this surprising result occurring by chance alone is _____ percent. If it is a small number, then it seems that the difference is genuine and unlikely to have occurred by chance. In fact, the probability of this result occurring, or something even more extreme, was calculated at that very number.

That number is the "p-value," and it has been customary in science to spend little time on studies that show results with p-values greater than 1 to 5 percent—that is, above 5 percent the findings are too likely to have occurred by chance alone. But p-values of less than 5 percent are more interesting, as they may represent a genuine phenomenon that was not included in our original understanding.

Lesson number nine: Find the p-value. An apparent observation with a p-value less than .05 suggests something is genuinely afoot and the apparent difference is not easily explained by chance.

Regression Analysis

We have observed that study design is strongest when all variables can be eliminated, reduced, or evenly distributed, but for the single variable in question. In those cases, we may be able to find a one-to-one correlation between the test variable and the outcome.

But what does one do when there are many—even hundreds—of variables in the equation so no such design is possible? In those cases, we turn to regression analysis: the study accounts for the many variables by mathe-matically equalizing all but one then seeing the degree to which the one remaining variable correlates to the outcome. For example, what predicts SAT scores—the SAT scores of the student's parents? The SAT scores of the other students in the school? The per-pupil expense spent on the student at the school? Absence of a felony conviction? Absence of becoming a parent prior to graduation? The possible factors correlating to SAT scores seem endless. So we test each one for its individual strength of correlation with the high SAT score. It is crucial, however, that we keep in mind regression does not show degree of *causality*, but degree of *correlation*.[6]

The overall approach in regression analysis is to make a best guess on a factor that you believe is powerfully associated with an outcome, collect data on all seemingly correlating factors, and describe the correlation between the one factor in question and the outcome as an approximate formula. You then measure how well your single variable predicts the observed outcomes and by how much you missed. The amount you missed, the "error," is a matter of both random chance and the sum of all other systematic effects (i.e., the other variables you've held as equal).

Those who are undaunted and want to include statistical analysis in their assessment of questions need to be comfortable with a couple of fig-ures in particular.

The squared correlation, R^2, is the proportion of the total spread accounted for by the variable you are studying ("the regression"). R^2 is always 0–1. For example, when $R^2 = 0.63$, then the proportion of the spread attributable to your variable of interest is 63 percent.

R^2 is interesting in itself, and it allows us to then focus on the remaining "error"—some of which is due to random chance, which is always present when we are using limited data to represent a general phenomenon. But some of it is due to the other variables that correlate with the outcome, but that are, for now, buried in the error. What is the relative contribution of each to the ultimate outcome studied? The goal then becomes to pull out of the "error" the influence of the other variables not yet studied: multiple regression analysis.

Another critical figure of which the numerate must not be afraid is R, which is, of course, the square root of R^2. R is called the "correlation coeffi-cient." A sign of "+" or "-" is assigned to represent whether increasing the variable increases the outcome or decreases it. A perfect positive correlation between the variable and the outcome results in a coefficient of +1, a perfect negative correlation in a coefficient of -1, and a total absence of correlation is a coefficient of 0. Intermediate values between +1 and 0 or -1 are the quantified expression of the degree of correlation.

Lesson number ten: Yes, one can calculate the degree to which each of many independent variables predicts an outcome. Likewise, when a person makes a claim of causation—"All you have to do in this country to be success-ful is to finish high school, not get married before you are eighteen, and not get pregnant before you are married."—you can put the claim to a valid test. To do so, one needs to know multiple regression analysis.

The search for valid evidence to inform important questions will often lead to scientific studies that involve statistical calculations and descriptions. It is a language that one must understand and be comfortable with if one is to comprehend the significance of the data and the relative strength of the findings. Learn the language. It's not as hard as you may think.[7]

Chapter Eight

Recognizing My Own Biases

In this book, we explored the question of how an individual can learn to think for himself or herself, deriving lessons from everyday debate, history, science, religion, rhetoric, human nature, and mathematics. In the end, we can think for ourselves only if we choose to. And if we choose to, then we have to accept the responsibility for the great amount of work, uncertainty, and humility (and the extraordinary amount of fun and discovery) that thinking for ourselves entails. We must:

- Be curious about what the better answers are.
- Admit that we already have a full set of established opinions, very few of which are based on an objective assessment of valid evidence.
- Be willing to sweep them all aside to start fresh.
- Cultivate the skills of research, of doing some fairly heavy lifting cognitively, with equal opportunity given to all opinions in any debate.
- Be ever mindful of one's own biases and managing them carefully so that they do not interfere with one's new and objective assessment.
- Allow ourselves to make timely decisions, even with full admissions of our irreducible uncertainty.
- Revisit the same questions to improve our answers continuously for our entire lives.

Faced with these challenges, it should not surprise us that so many people prefer the easy and consoling answers, feigned certainty, and reassuring myths. But we choose otherwise, wherever that choice takes us.

While this book has attempted to help the reader identify the strengths and weaknesses of arguments encountered, I suppose one could apply these rules to mischievous ends: employ the fallacies of reason rather than weed them out. As such, this book becomes just another Machiavellian guide for those who seek to persuade at any cost, to manipulate, or to deceive. Well, the streets and the markets have plenty of such teachers already, so I do not think my humble contribution would add much to that armamentarium. Rather, it is the *audience* of those schemers to whom I have addressed this book, those trusting and perhaps naïve enough to not realize that they are being bamboozled. Perhaps now they will know what to look out for. A sucker may be born every minute, but perhaps I can help convert a few into fair and reasonable skeptics.

However, if this book helps the reader to be capable only of criticizing the thoughts and words of others, then I and this book have failed. The real challenge is to critically scrutinize one's own thoughts and words, to be involved in the great questions of our day, and see if you can leave the world just a bit better for your contribution.

This great task never ends, because at any time, new data may arise that may require an honest reassessment of the entire issue and may even lead to a maddening series of reversals of our conclusion. If that is what the evidence demands, then so be it. Your less analytically minded friends may think you a lunatic, but by the time you have come this far down the road of reason, they probably will have decided so already. You can handle that.

Nonetheless, some great people are steadfastly unafraid to change, and on them the future depends.

*I am not at all concerned with appearing consistent. In my pursuit after truth
I have discarded many ideas and learnt many new things. Old as I am in age,
I have no feeling that I have ceased to grow inwardly or that my growth will
stop with the dissolution of my flesh. What I am concerned with is my readi-
ness to obey the call of Truth, my God, from moment to moment.*

—Mahatma Gandhi, Harijan, April 29, 1933,

All Men Are Brothers

Appendix

How to Teach Your Child to Think Irrationally (with Alternatives)

> *Impressions received in childhood cannot be erased from the soul.*
> —Frederick the Great, from Asprey's
> Frederick the Great—The Magnificent Enigma

> *The chief cause of our errors is to be found in the prejudices of our childhood ... principles of which I allowed myself in youth to be persuaded without having inquired into their truth.*
> —Rene Descartes, as quoted by Will Durant
> in The Story of Civilization

We adopt our parents' ways not only in the *content* of our beliefs but in the *way* we learned to think critically (or uncritically) to arrive at our beliefs. Part of the struggle of learning how to think critically is in coming to understand how we learned noncritical and irrational thinking in the first place. We become adept at critical thinking only if we are taught to do so and practice it in everyday life until it is natural and automatic. The same is true for irrational thinking. At an early age, it is taught to us and we practice it daily, and most of us never discover our error.

Questions, not answers, are the propelling force in the advancement of humanity. It was questions that drove the ancient Greeks to cultural heights never before seen on the planet. It was contentment with the given answers that caused Rome to rest on its laurels and to fail to advance the sciences. It was hostility to questions that buried rationality for forty-five generations of humanity, until Copernicus, Kepler, Newton, and many others courageously came forward with questions again. The questions these men asked were no different from those our children ask us: When the sun goes down, where does it go? What are stars? Why doesn't the moon fall? If the Earth is a ball, why don't the people underneath fall off?

We must have patience with the questions our children ask, as much as we wish authorities had had patience with those early scientists. When we are with our children, we discover how hard it can be to have patience with their questions, how exhausting it is to encourage and foster their questions and explorations, and, as they grow into adults, how much we recoil when they come up with answers different from the ones we cherish.

Here I explore a little more closely the accidental miseducation of our youth as it occurs with some of the classic questions and classic answers. It offers some suggestions for parents and future parents on how to teach and model a more rational way to think, decide, explain, and do.

The best time to teach children is during the routine events of everyday life. In fact, parents are teaching them constantly, whether the parents know it or not, because, for better or worse, children are constantly absorbing their parents' behaviors, speech, reactions, language, and modes of thinking. What parents have not at some time found themselves astonished and perhaps mortified at what the child has learned and demonstrated for friends and strangers?

Let us consider some of the common things that parents find themselves saying and teaching to their child.

Because I said so.
Lesson learned: Blind obedience to authority. Intimidation of the other person "wins" arguments.

Alternative: "Because Mommy and Daddy think it is important to eat dinner together. We think dinner is a good opportunity to talk to each other about the day, and it helps us get along better. We don't get to spend enough time together as it is."

Lesson learned: There are reasons for my requests. The reasons themselves are important teaching points.

Quit asking me "Why?" all the time!
Lesson learned: Being inquisitive is annoying and bad.

Alternative: "Because playing kickball with the other kids will help you be better at kickball. … Because if you get better at kickball, you'll realize you can practice and get better at other things too. … Because all your life you will encounter new things, and you can't be afraid to have difficulty while you are learning. … Because learning something new is often difficult. … Because learning is a kind of change and change is often difficult. … Because change requires mental and physical energy and often a lot of time. … Because time and energy are things most people feel they don't have enough of, and I'm running out of both right now. ... " Just keep answering honestly and accurately.

Lesson learned: Your questions are valid and deserve honest answers. The answers themselves teach a great deal. When the parent reaches exhaustion, it is okay to teach "limit setting" and to negotiate to defer continuation of the conversation until later in the day. This too is an important skill to teach.

Be quiet! The TV news is on!

Lessons learned: This daily event is more important than us learning something about each other through discussion and more important than learning how to learn. Also, the daily news is worth thirty minutes of your time.

Alternative: "Hey, that's a good question! Let's turn off the TV and talk about it." Or, "That's a great question. But I'd really like to hear what this man is saying, so is it okay if we discuss that after the news is over?"

It's blue because the sky is blue. Who am I—Carl Sagan?

Lessons learned: Tautological arguments (of self-justification) are valid and are effective ways to confuse a person into befuddled silence. Sarcasm is an effective tool to stifle tough questions.

Alternative: "See how this glass makes a spot on the wall with the colors of the rainbow? That's because sunlight really has a lot of colors mixed in. The air in the sky is like the glass, and at different times we see different colors." Or "Hmm, I really don't know. Let's see if we can find it in our encyclopedia. By the way, I don't have all the answers, no one does. But I can show you how to look for them. And I want you to admire people not because they seem to have all the answers, but because they show you how to find them."

Lesson learned: The true answer, if you know it. Or, that no one knows everything, and that is okay. We need to be comfortable saying we don't know; we need to know how to find answers. It is a family priority to have access to some resources that give us answers.

Bubbles went to goldfish heaven. He is very happy there, and you should be happy for him. You'll see him there one day.

Lesson learned: If you don't know the answer, make something up. Denial: death isn't really death, because you are alive somewhere else. Sad people make other people uncomfortable, so let's make up a reason to try to be happy when, in fact, we are very sad.

Alternative: "Remember the dead bird we saw in the woods? He was dissolving into little pieces and going into the soil. Then the grass and animals use the parts of him in the soil to be strong and healthy. Everything takes its turn in life and then dies to allow something else to have its turn to live. It's okay to be sad that Bubbles died; I am too. But after we are sad for a while it is important that we can be happy again about everything else in life, including the grass and trees that are healthy, in part, because of Bubbles."

Lesson learned: Death is a fact of life, and what those facts are. We can accept death graciously and without fear. It is okay to be sad for an appropriate time. Life doesn't give us everything we want.

Well, if he jumped off a cliff, would you do that too?

Actually, I like that answer, ergo the dangers of blind obedience. But I would clarify it a little by adding, "If Billy's parents think it's okay for him to go swimming at age six with no parents around, that's their business. But the reason I don't feel that way is that swimming can be dangerous, and you haven't shown me you can swim twenty-five meters in less than two minutes yet. That was our deal. Don't go along with what someone says or does just because you like other things he said or did. You have to decide for yourself each time something new comes up. It's okay to go along sometimes and not go along other times."

Danger: While the cliff analogy is fun to say, it teaches the rhetoric device of *ad absurdum*—of extrapolating to an absurd extreme and discussing that point instead of the point at hand. Going to the store with her friend and her friend's responsible mother is not anywhere near the same as jumping off a cliff, or even unsupervised swimming.

You didn't look like you were listening in church today. Shame on you.

Lesson learned: The important thing is to look like you're listening. Be a poseur. If your mind is wandering, you are a bad person.

Alternative: "You looked like you were thinking. What were you thinking about? Haha, well, you know what? Part of the time I was thinking about the office. But, truthfully, I don't think the minister expects us to hear every word; in fact, he's probably hoping that if he can just get us thinking about some of what he said, he'd be happy. Of course, we should make a good strong effort to think about the things he's talking about since we're there, and think about nuclear space monkeys or the office later. What did you think when he said … "

Lessons learned: It's hard for everyone to pay attention all the time. But it is a skill worth trying to cultivate. However, allowing an idea to stimulate other new ideas of your own and thinking about those can be very fruitful too. It helps to think about things if you can talk about them with other people.

Ghosts and monsters in the basement? There is no such thing. Now go get the screwdriver.
Lesson learned: You don't need to be persuasive, just firm. And the guy who told me he doesn't know everything is sending me to my death.

Alternative: Tell her why you know that there are no monsters or ghosts. "In all the world, no one has ever gotten a real picture of a monster or ghost. It's all make-believe because basements are actually as empty and boring as they look. We can go to any museum or zoo in the world, and I guarantee you there is not a single monster or ghost there—because there aren't any! People tell made-up stories because real life isn't always exciting enough, but it's important to know what's real and what's not."

Lesson learned: Ghosts and monsters aren't real. But there are real reasons why people like to pretend. Make-believe stories can be fun, even getting scared can be fun, but in the end we know that it's just make-believe, and we have to keep that separate in our minds from reality. There are specific ways to prove to oneself what the reality is.

No, I don't have any pictures of the spirit of Granddad, and the museums and zoos don't have monsters of hell, you're right. But they are real, it's true, so make Granddad happy and always tell the truth—or you might go to hell yourself. It's just a matter of faith.
Lesson learned: Don't try to understand, just go with it. If it makes no sense, call it "a matter of faith" and quit asking questions, or you are in serious trouble.

Alternative: Reference above.

Someone in your family stole the dollar I gave you?! That's stupid!
Lesson learned: I'm stupid.

Alternative: Occam's Razor. "Hmm … well, what are the possibilities? Someone stole it … or you misplaced it … or mischievous little gnomes hid it … or someone mistook it for their own. Okay, of those possibilities, which is most likely? On other occasions when you were missing something, what was the cause? Right, you forgot where you put it. Have we ever seen gnomes in this house? Do we even believe gnomes exist? Why is it tempting to say someone stole it? Right, because then it's not our fault, and we can say someone else is responsible for the hassle. Just because that's easier for us doesn't mean it's true, right? All right then, someone else mistaking it for their own and simply misplacing it are the two most likely explanations. Let's ask everyone in the house if they've seen it, then we'll start thinking about where you may have misplaced it."

Lesson learned: Don't react emotionally or be accusative, though that may be a first impulse. Approach the problem rationally. Consider the possibilities and decide which one is most likely. Figure out how to sort through them one at a time. Don't let your frustration trip you up or cause you to make unfounded accusations.

Why are the sofa cushions everywhere? So you *lost* the dollar I gave you?! Or did you spend it on candy?! Well, that's the last time I trust you with money!

Lesson learned—React emotionally, ad hominem, make assumptions, and be punitive.

Alternative: Teach calm, methodical problem solving and forgiveness. "Hmm … let's put all the sofa cushions back and stop and think about it. When was the last time you can remember having the dollar? What were you wearing at the time? Did you check the pockets of those clothes? Where have you been since then? Have you retraced your steps? Did you get distracted by anything? The phone? The dog? The bathroom?"

Lesson learned: While it is tempting to react physically (searching the house and pulling out cushions), it may save time and trouble to stop and think first. As above, approach problems calmly, methodically, and rationally.

We have meatloaf on Fridays because we always have meatloaf on Fridays.

Lesson learned: Tradition is self-justifying.

Alternative: "Hmm … I don't know. Maybe we're just in a routine and routines are comfortable." Or, "Because I really like meatloaf, but your mother would throw a fit if I made it twice every week." Or, "Because this is the sacred meatloaf, so we eat it to feel a connection to our religious community, past and present. When we eat it, there's nothing magical about the meatloaf per se, but I associate it with doing it every Friday with my own my parents, and that reminds me of all the truly important things they taught to me and that I want to teach you."

Lesson learned: Tradition is not self-justifying, but there may be valid reasons for traditions. Usually, the tradition itself is not the point, but the ideas we associate with it may be. If the reasons don't seem valid to you anymore or if the cost is too high, change. But change them with care.

I think you get the idea. Encourage questions. Be patient with the questions. Answer truthfully. Admit when you do not know the answer. Demonstrate how to find the answer. Demonstrate how to change your mind when a better answer comes along. Be careful with fables, as children might confuse them with literal truth. Explain that just because other people approach questions in different ways, we are very careful about interfering.

Think for yourself.

Endnotes

Acknowledgments

[1] The English physicist Sir Isaac Newton (1642–1727) once remarked, "If I have seen further, it is by standing on the shoulders of giants."

Introduction

[1] One of the reasons this is so difficult is that since childhood, we may have had bad habits that impair our ability to think critically and that are, unfortunately, deeply ingrained. If some of the tools presented in this book seem strange, or even counterintuitive, it might be because they conflict with your habits of doing things quite differently, which may or may not be promoting good analysis. If you suspect that this is the case, you might want to peek now at this book's last chapter. There I discuss the wrong ways and right ways to teach children to think analytically. Do you recognize any from your own childhood?

[2] The poem is actually the work of Father Gassalasca Jape, S. J., whose illustrative quotations add to both the substance and the satire of Bierce's original work. See the edition by Dover Publications, Inc., 1993.

[3] Apologies to Monty Python's Flying Circus, *Monty Python and the Holy Grail*.

[4] John Locke (1632–1704) was an English philosopher and statesman whose ideas about humanity's natural and civil behavior had a profound influence on the structure of the U.S. Constitution. The term "tabula rasa" has been criticized for not representing the vast store of animal instinct that we are born with, and this is a just criticism. Instincts of fight, flight, sex, possessiveness, and so on, are indeed powerful—but not always helpful. Regardless, for most complex social issues, we need to supplement, and perhaps moderate, our instincts greatly with a thorough education on history, science, psychology, morality, and economics.

[5] More on learning to live with imperfections can be found in Kenneth R. Hammond's *Human Judgment and Social Policy: Irreducible Uncertainty, Inevitable Error, Unavoidable Injustice* (New York: Oxford University Press, 1996).

[6] As with many noble and useful principles, it is a tempting error to render this

principle into an absolute. Who would tell a gravely ill five-year-old child who is enjoying her last thrilling visit from Santa Claus that it is all pretend? The principle of protecting the innocent or appeasing the dangerous with lies is valid in some cases—but in what cases? Do your neighbors have a right to information about your health and your sex life? Do your constituents? Under what circumstances? Does society have a right to "classified" national security secrets? Which ones? Should every useful myth be stripped from the faithful masses? Can the masses be trusted with the truth? In this book, discussion of truth, facts, and reality shall be primarily in the context of understanding science, history, psychology, and other realities or fictions of the larger world as they pertain to weighing arguments and making complex decisions.

7. Some readers will see the ghosts of certain philosophers flitting through these pages; the ghosts are real. If I do not summon them up more explicitly, it is partly to keep this book from being weighed down with what my wife would call "academic name-dropping." Also, it is not the philosophers who are most important; it is the concepts they addressed, which this book presents more fully. The books to which I will frequently refer, however, flesh out the philosophers, and I sincerely hope the reader will go to and thoroughly study those books next.

Chapter One

1. Another book about dead white guys? For a country that can't get enough take-out food from every ethnic group on Earth, I am amazed that some of us Americans so often slap our hands over our ears and refuse to take any ideas from a group because of the race of the persons presenting them. Ideas are to be rejected or embraced by their inherent merit, not by whether we have a sense of cultural familiarity or identity associated with them. Ideas—whether "Western" or not in origin—are universal: they belong equally to any and every race or epoch that cares to consider and debate them. If we regress into arguments of who came up with certain ideas first, or if we shrink from some ideas because we do not consider them "ours" or embrace too closely the ideas that we first proposed, then we limit dramatically our access to the best ideas from every corner of the globe.

Most of the people throughout the world who think scientifically are certainly not Greek, after all. And the Greeks and the northern European barbarians-turned-scientists can hardly claim any special genius as their lineage. They, as well as all other humans of whatever race, had required a few million years to come up with the scientific method. When the scientific method emerged and then reached fullness during the enlightenment, it was possible only through innumerable contributions from non-Western cultures. Furthermore, there is much to say about the attempts, successes, and failures of non-Western cultures to use reason, to lay bare the general principles of rational thought, to codify it as a scientific method, and to stand firmly against irrationality and superstition. For the purposes of this book, however, the lessons primarily from Western history will illustrate the point; any reader who knows of parallel cultural movements, good and bad, in other cultures is encouraged to contact the author at thinkforyourself@wispertel.net.

[2.] The notion of the Cosmic Calendar is Carl Sagan's. It is represented not in his book *Cosmos* (New York: Ballantine Books, 1985), but in the video series of the same name (*Cosmos* by Carl Sagan Productions, Inc., and TBS Productions, Inc., 1989). No claim at precision is made here, as the quality of the data for each of these events is variable, but it is the general scheme that represents our place in time.

[3.] Charles Singer, *A Short History of Science to the Nineteenth Century* (Mineola, N.Y.: Dover Publications, 1997), 5.

[4.] Refer to the bibliography regarding *The Story of Civilization, The Discoverers: An Illustrated History of Man's Search to Know His World and Himself,* and *Cosmos.*

[5.] Adam and Eve did not enjoy the only sacred grove—the Greeks also had theirs, where the Golden Fleece was kept on a tree, guarded by a snake or dragon. The Epic of Gilgamesh has numerous stories of striking parallel to those of the Hebrews, including such features as a deceitful serpent and a great flood.

[6.] Ibid.

[7.] Will Durant, *The Story of Civilization, Volume II: The Life of Greece* (New York: Simon and Schuster, Inc., 1966), 340.

[8.] Ibid., 344.

[9.] Religions of the Mediterranean often included a "Son of God," who was usually considered part human and part divine, as they were the offspring of a God, the Father, and a mortal woman. In Greek religion alone, which by 300 B.C.E. had substantial influence on the characters and traditional stories of Roman religion, God, the Father, Zeus impregnated Demeter to produce Persophone, Dione to produce Aphrodite, Leto to produce Apollo and Artemis, and Maia to produce Hermes, as well as others.

Augustus was not the only historical person thought to be a Son of God. For example, Alexander the Great visited the Oracle of Zeus and was hailed as the Son of God; his reputation as divine deliverer from Greek dominance grew, and he may have even begun to believe it himself (see Jeremy McInerney, University of Pennsylvania, *Ancient Greek Civilizations* [Chantilly, Va.: *The Great Courses* by The Teaching Company], DVD). Apollonius of Tyana was born in the first few years of the first century A.D.; his conception was announced to his mother by a divine message calling him Son of God. He was born of miraculous signs, came to be regarded as a youth prodigy among the wise men of the community, grew to become a traveling preacher of the pious and ascetic life, performed miracles, was considered to be both human and divine, attracted several disciples, and, due to his growing popularity, eventually came in conflict with the Roman authorities, who imprisoned him. See Bart D. Ehrman, University of North Carolina at Chapel Hill, *The New Testament* (Chantilly, Va.: *The Great Courses* by The Teaching Company), DVD.

[10.] Bart D. Ehrman, University of North Carolina at Chapel Hill, *Lost Christianities* (Chantilly, Va.: *The Great Courses* by The Teaching Company), DVD.

[11.] Will Durant, *The Story of Civilization, Volume III: Caesar and Christ* (New York: Simon and Schuster, Inc., 1980).

[12.] Microsoft Corporation, *2001 Microsoft® Encarta® Encyclopedia Deluxe*, copyright © Microsoft Corporation.

[13.] De Lacy O'Leary, *Arabic Thought and Its Place in History* (Mineola, N.Y.: Dover Publications, Inc., 2003), 168–180, 219–225.

[14.] Will Durant, *The Story of Civilization, Volume IV: The Age of Faith* (New York: Simon and Schuster, Inc. 1950), 241.

[15.] Microsoft Corporation, *2001 Microsoft® Encarta® Encyclopedia Deluxe*, copyright © Microsoft Corporation.

[16.] Will Durant, *The Story of Civilization, Volume IV: The Age of Faith* (New York: Simon and Schuster, 1950), 240.

[17.] Muzaffar, Iqbal, *Islam and Science* (Ashgate Science and Religion Series) (Burlington, Vt.: Ashgate Publishing Co., 2002).

[18.] Will Durant, *The Story of Civilization, Volume IV: The Age of Faith* (New York: Simon and Schuster, Inc., 1950), 243.

[19.] De Lacy O'Leary, *Arabic Thought and Its Place in History* (Mineola, N.Y.: Dover Publications, Inc., 2003), 168–180, 219–225.

[20.] Microsoft Corporation, *2001 Microsoft® Encarta® Encyclopedia Deluxe*, copyright © Microsoft Corporation.

[21.] For one author's perspective on this history, see chapter five of Muzaffar Iqbal's *Islam and Science* (Ashgate Science and Religion Series). Iqbal's thesis is that there is a rich and creative history of science in Islam that reconciles scientific thinking and the universal truth contained in the Koran. His understanding of science may, however, be different from that of Western scientists. He also finds many other reasons for the decline of science in Islam: he rejects the widely held notion that religious orthodoxy was (and is) repressive to free scientific thought and instead cites Western racism, denial of Islamic scientists' contributions, the Mongol invasion, economic and political distractions, neglect (although not repression) of science, and other causes.

22. Microsoft Corporation, *2001 Microsoft® Encarta® Encyclopedia Deluxe*, copyright © Microsoft Corporation.

23. Bernard Lewis, *What Went Wrong? The Clash between Islam and Modernity in the Middle East* (New York: Oxford University Press, Perennial Edition, 2003), 80.

24. Scrutiny of Deepak Chopra's "New" Age literature of the 1990s reveals well-selling and spectacularly vacuous pseudoscience. The reader is referred to Chopra's various works and also to Wendy Kaminer's comment on them in *Sleeping with Extra-Terrestrials: The Rise of Irrationalism and Perils of Piety* (New York: Pantheon Books, 1999).

25. Walters, Kerry S., *The American Deists: Voices of Reason and Dissent in the Early Republic* (Lawrence, Kans.: University Press of Kansas, 1992).

26. The timeline I have represented here takes us only to the founding of the United States, plus a few monumental events thereafter. But the marvelous history of reason continues on, though I do not go into it in this book. For an elegant and engaging review of the role of secular thought in the United States, I refer the reader to the Susan Jacoby's *Freethinkers: A History of American Secularism* (New York: Henry Holt and Company, 2004).

27. Karl Ernst von Baer (1792–1876), an Estonian naturalist and biologist, set forth the notion that individuals (ontogeny) must retrace the developmental steps of the species (phylogeny). Consider the story of a journey from a single-cell organism to a complex, multicellular organism rooted to an aquatic floor; to an aquatic organism with a tail; to a tailed animal that moves then onto land and breathes air; then walks on four limbs; then stands erect; then uses tools; develops language; and, in its most recent stage, learns to reason systematically. This is certainly a story well familiar to both embryologists and paleoanthropologists. The process was an evolutionary path originally cut by many generations, each adding a new genetic innovation as another step. The individual, while enjoying some time-saving shortcuts, must make each new change personally.

There may be a similar correlation between the process of intellectual development that humanity has experienced and that which the individual retraces: we see stages dominated by instinct, family, tribe, violence, certainty, and myth all aimed primarily at survival and security slowly giving way to contemplation, empathy, collaboration, knowledgeable uncertainty, and discovery of truth as security is assured and the mind freed. In this case, however, the individual of today can choose to stop development at any point, declining to make the next change, never achieving systematic use of reason, and, despite outward appearances of language and dress, intellectually residing at a mind-set characteristic more of an era in human history centuries or millennia past. Progress in that journey of intellectual development we call education; some books and teachers may offer experiences in which grand ideas of the centuries are quickly communicated—or not. The educational journey is a phenomenon quite separate from learning a trade.

Chapter Two

1. Of all the works cited in this book, one of my strongest recommendations is for Richard Hofstadter's *Anti-Intellectualism in American Life* (New York: Vintage Books, 1963), winner of the Pulitzer Prize in 1964.

2. Caveat: I think that the most dogmatic zealots are often the most interesting. In discussions with them, I have no expectation of a mutual exchange and consideration of ideas, but I do find them an efficient resource by which I can learn all the arguments from one side of an issue. I learn a lot, about both the merits and the limitations of their argument, but the analysis doesn't always feel like the collaboration it should be.

3. Stanley Appelbaum, ed., *Confucius: The Analects*, vol. I, book III, chapt. XVII (New York: Denver Publications Inc. 1995).

4. Sometimes our response to a problem is as unfounded as our explanation of it. Doing a rain dance, taking an antibiotic for a viral condition, and Keynesian economics may all be exercises that give us the satisfaction of having done *something* to affect a change, even though there is little evidence it will help and perhaps plenty of evidence that we are deluding ourselves. It may be contrary to human nature to do nothing, but it is certainly contrary to good sense and human progress to persist with a futile intervention rather than to admit failure and move on to discover an effective one.

5. Michael Abrams, "Sight Unseen," *Discover*, June 2002, Vol. 23, No. 6.

6. This is not a perfect analogy, and perhaps the algebraic "x" as placeholder for the unknown quantity would be a more precise analogy, but the history of struggling without the "x" is less well-known to historians—or at least to this one. See *The Nothing That Is: A Natural History of Zero* by Ellen Kaplan and Robert Kaplan (New York: Oxford University Press, 1999).

7. As some of my students would say in response to an interesting disagreement, "Hey (shrug), it's all good." This vapid nonassessment is probably the surest way to make me pass out in front of the class.

8. Quoted indirectly from Tom Hirschfield, *Business Dad* (Boston: Little Brown and Company, 1999), 185.

9. This example serves to make the point on more than one dimension because it

may be based on fiction or, to be generous, on unusually fine distinctions. In 1986, Laura Martin reported in *American Anthropologist* that previous anthropologists were mistaken on this matter and that the apparent variety of words was really just a body of variations created with suffixes and root words. Thus, it may well be that the anthropologists were seeing finer distinctions of words than the Inuit were seeing types of snow. Thanks to Frances E. Mascia-Lees, editor of *American Anthropology*, in "The Chronicle of Higher Education," *Chronicle Review*, May 31, 2002.

I was tempted to drop this whole example of the Inuit and say something about my seeing a few types of pine trees while dendrologists see hundreds of species of gymnosperms, but the Inuit story is too much fun to leave out. The fun-factor is another important consideration in the perpetuation of myths, but more of that later.

10. This tendency can be overcome, as is sometimes the case in heroic self-sacrifice for the "other" group. Some may argue, however, that the desire to achieve public acclaim and to avoid public scorn is often the underlying self-interest that can be powerful enough to outweigh any loss, even loss of one's own life.

For a fascinating look at apparently altruistic behavior that may be rooted in a self-interest of sorts, see *Ethology: The Mechanisms and Evolution of Behavior* by James L. Gould (New York: W. W. Norton and Company, 1982).

11. As much as that last bit sounds like a veer into a fantasy novel, it is actually the proper terminology for the routine and predictable life cycle of stars, including our own.

12. 1. Conditional 2. Unconditional 3. Unconditional 4. Conditional

13. These can be found in their entirety at www.papalencyclicals.net/all.htm.

14. A brief comment is necessary to clarify the area between these extremes of conditional and unconditional faith. Between the cautious and conditional pragmatists versus the certain and unconditional believers are those who are the high-stakes gamblers: the ones who "have faith" that something is real or can be done despite full knowledge of the long odds. These are the people who believed that man could land on the moon and return safely, or who believed the Allies could surprise and attack Hitler by landing 175,000 troops on the coast of Normandy in one day, for example. It is the glory of these people who defy all warnings and realize the "impossible" that make it tempting for others to believe against the odds. But such odds are usually beaten by very hard, rational pragmatism that prepares the heroes for their day. If admirers come along behind and think they too can beat the odds and earn the glory but intend to do it "on faith" only, without also having a solid real-world basis for their gamble, then they are likely to have different and disappointing

results. In the skills of reason, people have the responsibility to know the real difference between a long shot and an unsupportable delusion and the distinction must be made based on legitimate evidence, not preference and wishful thinking.

15. Tami Silicio, a fifty-year-old cargo worker whose photograph of flag-draped coffins bearing the remains of U.S. soldiers was published on the front page of the *Seattle Times*, was fired April 21, 2004, by Maytag Aircraft Corporation after military officials raised "very specific concerns" related to the photograph. Her husband, David Landry, also an employee of Maytag, was also fired. William L. Silva, Maytag president, said Silicio violated company and federal government rules (Hal Bernton and Ray Rivera, "How Two Women, One Photo, Stirred National Debate," *Seattle Times*, Sunday, April 25, 2004, Associated Press).

16. Bernard Lewis, *Emergence of Modern Turkey: Studies in Middle Eastern History* (New York: Oxford University Press, 2001), 108.

17. The Second Amendment reads, "A well regulated Militia, being necessary to the security of a free State, the right of the people to keep and bear Arms, shall not be infringed." While a literal interpretation suggests that "well regulated" opens the door to gun control—especially in a modern country of hundreds of millions of guns, many of them automatic or semi-automatic weapons, and 30,000-plus gun deaths per year—but other scholars interpret it as conferring an individual rights to a gun as found in the latter half of the sentence.

18. Statisticians tell us that when we make a positive decision ("guilty" or "invade") in error, that is a "Type I error"; the opposite is a "Type II error." My brain seems incapable of remembering the correct name for each error. I am sure I will live my entire life still misnaming them and not even knowing each time I am doing so if I am making a Type I or a Type II error.

19. A thorough, analytic, and relevant description of this dynamic between information, judgment, and error is in Kenneth R. Hammond's *Human Judgment and Social Policy: Irreducible Uncertainty, Inevitable Error, Unavoidable Injustice* (New York: Oxford University Press, 1996).

Chapter Three

1. See *The Great Betrayal: Fraud in Science* by Horace Freeland Judson (New York: Harcourt, 2004). Also *The Baltimore Case: A Trial of Politics, Science, and Character* by D. J. Kevles (New York: W. W. Norton, 1982).

2. The CAGE mnemonic is based on J. A. Ewing, "Detecting Alcoholism: The CAGE Questionnaire," *Journal of the American Medical Association* 252, no. 14 (12 October 1984): 1905–1907.

[3] John Macleod, et. al., "Psychological and Social Sequella of Cannabis and Other Illicit Drug Use by Young People: A Systematic Review of Longitudinal, General Population Studies." *The Lancet* 363 (15 May 2004): 1579–1588.

[4] Perhaps there is an element of confounding factors here too. It may be that persons averse to getting a needle in their arm delay getting their vaccine until the sheer number of persons around them with the flu convinces them they should go ahead with the shot. It takes a week or more for the shot to create immunity, but by that time, the many exposures have already led to infection, and a few days after the shot, the person has flu symptoms.

[5] Hill, A. B. *The Environment and Disease: Association or Causation?* vol. 58 of *Proceedings of the Royal Society of Medicine* (London, 1965), 295–300.

[6] File this phrase along with other dubious reassurances: "If we do this, it won't change the way I feel about you," "The check is in the mail," and "I won't get mad—just tell me the truth." Remember, sometimes it's true.

Chapter Four

[1] Will Durant, *The Story of Civilization, Volume II: The Life of Greece* (New York: Simon and Schuster, Inc., 1966), 345.

[2] Gamow was the first to propose the scientifically based general theory of the phenomenon of a titanic explosion at the start of the observable universe. When British astronomer Fred Hoyle disparagingly characterized the notion as a "big bang," the theory found its name.

[3] Hall, N. K., R. R. Terry, J. T. Crosby, et. al. *Pain*. Monograph Edition No. 275, Home Study Self-Assessment Program (Leawood, Kan.:, American Academy of Family Physicians, April 2002), 16.

[4] See also Charles Singer, *A Short History of Science*, 29.

[5] Cited by Sagan, *Demon Haunted World*, 119.

[6] William Manchester, *A World Lit Only by Fire: The Medieval Mind and the Renaissance—Portrait of an Age* (New York: Back Bay Books, 1993), 14.

[7] Microsoft Corporation, *2001 Microsoft® Encarta® Encyclopedia Deluxe*, copyright © Microsoft Corporation.

[8] I emphasize "most." I continue to search the Koran for something that allows for a chance that there is much knowledge outside the Koran, or that individual judgment apart from the teachings of the Koran may at times be acceptable. I have not found it. I have, however, found "Believe ye then part of the Book and deny part? But what shall be the meed of him among you who doth this, but shame in this life?" (Sura 2, v. 79). Similarly, Sura 10:16 I find discouraging. There may be some consolation in Sura 17:37, "Give full measure when you measure, and weigh with a just balance."

⁹· Henry David Thoreau, *Walden and Civil Disobedience* (New York: Panguin Books, 1986), 4II. *Civil Disobedience* was first published in 1849.

¹⁰· Will Durant, *The Story of Civilization, Volume II: The Life of Greece* (New York: Simon and Schuster, 1966), 347.

¹¹· George Sinkler, *The Racial Attitudes of American Presidents from Lincoln to Roosevelt* (Garden City, N.Y.: Doubleday/Anchor Books, 1972). Cited in Christopher Cerf and Victor Navasky, *The Experts Speak* (New York: Villard Publishers, 1998), 22.

¹²· D. G. Tendulkar, *Mahatma: Life of Mohandas Karamchand Gandhi* vol. IV (Publishers Jhaveri and D. G. Tendulkar, 1952), 73.

¹³· Karen Wright, "Einstein's Greatest Mistake" in "The Master's Mistakes," *Discover* 25, no. 9 (September 2004). Wright explains that the lambda symbol, λ, in the second term from the left, known as the Cosmological Constant, was inserted by Einstein as a fudge factor so that the equation would describe a static universe rather than an expanding or contracting universe, which the equation otherwise supported. In 1929, Edwin Hubble observed the Doppler Effect for light from various stars and discovered that the universe is, in fact, expanding.

Chapter Five

¹· Incidentally, this word itself has an ignominious history, equating the symbol of the female, the uterus, in Greek *hysterikós*, as in "hysterectomy," with physical and mental uncontrollability.

²· Referring to the events in Beslan, North Ossetia, Russia, on September 3, 2004.

³· Despite a handful of contacts between Saddam Hussein and al-Qaeda, the 9/II Commission found no credible evidence of collaboration, according to *The 9/II Commission Report: Final Report of the National Commission on Terrorist Attacks Upon the United States* (New York: W. W. Norton and Company, 2004). Bush's response and his immediate follow-up comments explaining the apparent discrepancy between strong statements claiming alliance against the United States (at a political rally in Dearborn, Michigan, October 2002, and by Vice President Dick Cheney just days before the 9/II report was released, for example), as well as the subsequent lack of evidence, might also be described as weasel words, semantics, or other types of misleading rhetoric. The reader is referred to those sections of this book.

⁴· On August 17, 1998, during Clinton's grand jury testimony before the independent counsel, the attorney asked, "Whether or not Mr. Bennett knew of your relationship with Ms. Lewinsky, the statement that there was 'no sex of any kind in any manner, shape, or form, with President Clinton' was an utterly false

statement. Is that correct?" To which Clinton answered, "It depends on what the meaning of the word 'is' is. If the—if he—if 'is' means 'is and never has been and is not'—that is one thing. If it means 'there is none,' that was a completely true statement."

5. Will Durant, *The Story of Civilization, Volume II: The Life of Greece* (New York: Simon and Schuster, 1966), 526.

Chapter Six

1. Existentialists may disagree with this vigorously. Existential philosophy characteristically refuses to allow any excuses as valid substitutes for reason. Every decision, and perhaps every emotion, is a personal choice for which the chooser is responsible. There is no hiding behind our lot in life, for we are all responsible for laboring to rise above it. There is no hiding behind historical injustice, for we cannot change the past; we can only learn from it, seek justice, and be responsible for ourselves now. There is no hiding behind God or any other authority who has told us what to think, for we freely choose to submit to that comfortable bondage. There is no allowance for fear, for it is a responsibility of adulthood that we face down fears. There is no waiver for indulging in appealing benefits of irrationality, for an inauthentic life is too high a price to pay. For an engaging and illuminating review of existentialism, I recommend the works of Robert C. Solomon, from the University of Texas.

2. Cadmus, in Greek mythology, Phoenician prince who founded the city of Thebes in Greece. When his sister Europa was kidnapped by the god Zeus, Cadmus was ordered by his father, the king of Phoenicia, to find her or not to return home. Unable to locate his sister, he consulted the Oracle of Delphi and was instructed to abandon his search and instead to found a city. Upon leaving Delphi, the Oracle advised, Cadmus would come upon a heifer, follow her, and build the city where she lay down to rest.

Near the site of the new city, Cadmus and his companions found a sacred grove guarded by a dragon. After the beast killed his companions, Cadmus slew the dragon and, on the advice of the goddess Athena, planted its teeth in the ground. Armed men sprang from the teeth and fought each other until all but five were killed. Cadmus enlisted the help of the victors in founding the citadel of the new city of Thebes, and they became the heads of its noble families. (Microsoft Corporation, *2001 Microsoft® Encarta® Encyclopedia Deluxe*, copyright © Microsoft Corporation.)

3. The infamous words of President William J. Clinton, January 20, 1998, on the *NewsHour* with Jim Lehrer.

4. It is difficult to overstate the importance of John Locke, an English statesman,

philosopher, and scientist, in the development of political thought in the American colonies. An advocate not only for property rights and separation of church and state, he was a powerful spokesman of freedom of speech and of the obligation for persons and institutions to respond with respect to those who would exercise their right to openly question and dissent.

[5] As referred to in Baron D'Holbach's 1772 work, *Common Sense, or Natural Ideas Opposed to Supernatural.*

Chapter Seven

[1] Okay, it does get more technical than that. Conceptually, the standard deviation is simply the average difference between each data point, x, and the mean of them, x̄. That would be x-x̄ for every data point, added up and divided by the total number of data points. We call that simple calculation for standard deviation "s." Very easy. But to compensate for some outliers screwing up the figures, we modify that slightly: we square each difference $(x-\bar{x})^2$, average those, and then take the square root of that average. Same idea as "s," but with this extra maneuver for accuracy we call it "σ" instead. Don't panic. Again, it's easier than it looks, and all the cool symbols are very satisfying. Statisticians have formulas that are really varieties of a theme and use different nomenclature in each variety. But once you get the basic concept, the scary-looking nomenclature falls into place. So once you know what an x means, then a μ is easy. When you grasp s, then σ makes sense. When you know what a z-point is, then a t-point is obvious. The statistically uninitiated will hear this with skepticism, but those who have taken Statistics 101 know the secret. Statistics books are the place to go for the details.

[2] This is a significant caveat. Many phenomena do not occur in a symmetrical distribution, often because of a natural limitation to the data. The number of minutes spent seeing a physician, for example, cannot be less than zero; the distribution may have an arithmetical mean around 15 minutes, but individual events may be far larger than that in rare circumstances.

[3] How do I know that? Because a derivation of the formula for standard deviation, as is applicable to this situation, is

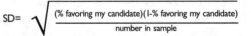

$$SD = \sqrt{\frac{(\% \text{ favoring my candidate})(1-\% \text{ favoring my candidate})}{\text{number in sample}}}$$

And we know that 95 percent of any such samples will fall within 2 SDs of the true figure. In this case, the percent favoring my candidate = 0.463, and the number in sample = 1,000, so SD calculates to 0.0157. That means that the true percentage of voters intending to vote for my candidate is, with a 95 percent probability, 46.3% +/- (1.57% x 2); that is, between 43.16% and 49.44%. Cool, huh? The formula can be adjusted for special circumstances.

4. U.S. Preventative Services Task Force, *Guide to Clinical Preventative Services, Second Edition*, 1995. (See also the *Journal of the American Medical Association*, 274, no. 22 (1995): 1800–1804.)

5. http://www.guttmacher.org/pubs/sfaa.html. AGI, State Facts about Abortion, 2003. Accessed April 14, 2004.

6. An enjoyable and fascinating look at how econometric analysis, including regression analysis, can be applied to everyday questions and serious public policy problems is in *Freakonomics: A Rogue Economist Explores the Hidden Side of Everything* by Steven D. Levitt and Stephen J. Dubner (New York: HarperCollins, 2005).

7. There are many wonderful books that describe statistics. My favorites are *The Cartoon Guide to Statistics* by Larry Gonick and Woollcott Smith (New York: HarperCollins, 1993) and *Statistical Techniques in Business and Economics, Tenth Edition* by Robert Mason, et. al. (New York: McGraw-Hill/Irwin, 1999).

Bibliography

Bierce, Ambrose, *The Devil's Dictionary* (New York: Dover Publications, Inc., 1993).

Boorstin, Daniel J., *The Discoverers: An Illustrated History of Man's Search to Know His World and Himself* (New York: Vintage Books, 1993).

Cerf, Christopher, and Victor Navasky, *The Experts Speak: The Definitive Compendium of Authoritative Misinformation, Revised Edition* (New York: Villard, 1984).

De Beauvoir, Simone, *The Second Sex*, translated and edited by H. M. Parshley (New York: Vintage Books, originally published in 1953, this edition, 1989).

Dodds, E. R., *The Greeks and the Irrational* (Boston: Beacon Press, 1957).

Dostoyevsky, Fyodor, *Crime and Punishment*, translated by Constance Garnett (New York: Bantam Skylark, first published in 1866, this translation, 1958).

Dostoyevsky, Fyodor, *The Brothers Karamazov*, translated by Constance Garnet, abridged by Edmund Fuller (New York: Dell Publishing Co., Inc., 1956).

Durant, Will and Ariel Durant, *The Story of Civilization, Volumes I–X* (New York: Simon & Schuster, 1935–1969).

Ehrman, Bart D., *Lost Christianities: The Battle for Scripture and the Faiths We Never Knew*, (New York: Oxford University Press, 2003).

Emerson, Ralph Waldo, "Self Reliance," first published in 1841, referred to here in *Selected Essays* (New York: Penguin Classics, 1982).

Encarta® Encyclopedia Deluxe 2001, copyright © Microsoft Corporation, 2001.

Engel, S. Morris, *With Good Reason: An Introduction to Informal Fallacies, Fifth Edition* (New York: St. Martin's Press, 1994).

Firth, Raymond, *Religion: A Humanist Interpretation* (London: Routledge, 1996).

Gandhi, Mahatma, *Harijan* magazine, April 29, 1933, as cited in *All Men Are Brothers: Autobiographical Reflections*, compiled and edited by Krishna Kripalani (New York: Continuum Publishing Company, 1990).

Gibbon, Edward, *The Decline and Fall of the Roman Empire*, originally published in London, 1787 (New York: Penguin, 1952).

Gould, James, *Ethology: The Mechanisms and Evolution of Behavior* (New York: W. W. Norton & Co., 1982).

Gould, Stephen Jay, *The Mismeasure of Man* (New York: W. W. Norton and Co., 1996).

Grant, Michael, *The Ancient Mediterranean* (New York: The Penguin Group, 1969).

Halpern, Diane F., *Thought and Knowledge, Third Edition* (Mahwah, N.J.: Lawrence Erlbaum Associates, 1996).

Hammond, Kenneth R., *Human Judgment and Social Policy: Irreducible Uncertainty, Inevitable Error, Unavoidable Injustice* (New York: Oxford University Press, 1996).

Hancock, Graham, *Lords of Poverty* (London: Macmillan London Ltd., 1989).

Hofstadter, Richard, *Anti-Intellectualism in American Life* (New York: Vintage Books, 1952).

Irving, John, *The Cider House Rules* (New York: Ballantine Books, 1985).

Jacoby, Susan. *Freethinkers: A History of American Secularism* (New York: Henry Holt and Company, 2004).

Judson, Horace Freeland. *The Great Betrayal: Fraud in Science* (New York: Harcourt, 2004).

Kaplan, Robert, *The Nothing That Is: A Natural History of Zero* (New York: Oxford University Press, 1999).

Kramnick, Isaac. *The Portable Enlightenment Reader* (New York: Penguin Books, 1995).

Levitt, Steven D. and Stephen J. Dubner. *Freakonomics: A Rogue Economist Explores the Hidden Side of Everything* (New York: Harper Collins, 2005).

Lewis, Bernard. *What Went Wrong? The Clash between Islam and Modernity in the Middle East* (New York: Oxford University Press, Perennial Edition, 2003).

Manchester, William, *A World Lit Only by Fire: The Medieval Mind and the Renaissance Portrait of an Age* (Boston: Little Brown, and Co., 1992).

Mander, Alfred, *Logic for the Millions* (New York: Philosophical Library, Inc., 1947).

Markem, Beryl, *West with the Night* (Stanford, Calif.: North Point Press, 1982).

Rossiter, Clinton, editor. *The Federalist Papers—Hamilton, Madison, Jay* (New York: The Penguin Group, 1961).

Sagan, Carl, *Cosmos* (New York: Random House, 1980).

Sagan, Carl, *The Demon-Haunted World: Science as a Candle in the Dark* (New York: Ballantine Books, 1996).

Sharples, R. W., *Stoics, Epicureans, and Sceptics: An Introduction to Hellinistic Philosophy* (London: Routledge, 1996).

Shermer, Michael, *Why People Believe Weird Things* (New York: W. H. Freeman and Company, 1997).

Singer, Barry and George O. Abell, *Science and the Paranormal: Probing the Existence of the Supernatural* (New York: Scribner's, 1981).

Singer, Charles, *A Short History of Science to the Nineteenth Century* (originally published by Oxford University Press, 1941, New York: Dover Publications, 1997).

Taylor, C. C. W., R. M. Hare, and Jonathan Barnes, *Greek Philosophers: Socrates, Plato, and Aristotle* (Berkshire, England: Oxford University Press, 1999).

Twain, Mark, *Life on the Mississippi* (originally published by Osgood in 1883, New York: Penguin Group, 1984).

Walters, Kerry. *The American Deists: Voices of Reason and Dissent in the Early Republic* (Lawrence: University Press of Kansas, 1992).